Nuclear Matter Theory

Nuclear Matter Theory

Omar Benhar
Stefano Fantoni

CRC Press is an imprint of the
Taylor & Francis Group, an informa business

CRC Press
Taylor & Francis Group
6000 Broken Sound Parkway NW, Suite 300
Boca Raton, FL 33487-2742

First issued in paperback 2021

ISBN 13: 978-1-03-223975-0 (pbk)
ISBN 13: 978-0-8153-8666-7 (hbk)

DOI: 10.1201/9781351175340

Library of Congress Cataloging-in-Publication Data

Names: Benhar, Omar, author. | Fantoni, S. (Stefano), author.
Title: Nuclear matter theory / Omar Benhar, Stefano Fantoni.
Description: Boca Raton : CRC Press, 2020. | Includes bibliographical references and index.
Identifiers: LCCN 2019048245 | ISBN 9780815386667 (hardback) | ISBN 9781351175340 (ebook)
Subjects: LCSH: Nuclear matter.
Classification: LCC QC793.3.N8 B46 2020 | DDC 539.7--dc23
LC record available at https://lccn.loc.gov/2019048245

Visit the Taylor & Francis Web site at
http://www.taylorandfrancis.com

and the CRC Press Web site at
http://www.crcpress.com

Contents

Preface

Nuclear matter can be thought of as a giant nucleus, consisting of an infinite number of protons and neutrons subject to strong interactions only. Theoretical studies of such a system, which greatly benefit from the simplifications granted by translation invariance, are a necessary intermediate step towards the description of atomic nuclei, and provide the basis for the development of accurate models of matter in the interior of compact stars.

While being a very lively research field, and the subject of a large number of original papers every year, nuclear matter theory—which lies at the interface of Nuclear Physics and the Physics of Quantum Fluids—has been seldom discussed in books, and never in a systematic and comprehensive fashion. In Nuclear Physics textbooks, nuclear matter is typically confined to one chapter at most, while monographs on Quantum Fluids fail to give proper emphasis to the complexity of nuclear dynamics.

The systematics of the nuclear charge-density distributions clearly indicates that interactions between protons and neutrons are strongly repulsive at short distance, and cannot be treated in perturbation theory using the basis of eigenstates of the non interacting system. Moreover, they exhibit a strong dependence on the total spin and isospin of the interacting particles, S and T, which entails a complex operator structure of the nuclear wave function. The very fact that a two-nucleon bound state is only observed with total spin and ispospin $S = 1$ and $T = 0$—the nucleus of ^{2}H, or deuteron—signals a significant spin-isospin dependence of the interaction.

In spite of the fact that the description of nuclear matter is fundamental in many areas of nuclear physics and astrophysics, doctoral students, as well as young researchers and senior scholars approaching this subject, have to resort largely to technical papers, or depend on the help of more learned colleagues, a problem that was made all the more severe by the groundbreaking progress of the past two decades. This book, providing a concise but exhaustive account of nuclear matter theory, from the early approaches to the most advanced developments, is meant to fill an empty spot in the existing literature.

In an effort to keep the book as self-contained as possible, we have included an introductory discussion of the models of nuclear dynamics and of the basic concepts of many-body theory. The different theoretical approaches to the nuclear many-body problem are analysed following their historical development, with an emphasis on the models that have been more widely applied to study the properties of nuclear matter.

Special attention is given to the recent applications of nuclear matter theory to the description of neutron star properties. A prominent role, in this context, is played by the studies of gravitational-wave emission from neutron stars, whose results will be of paramount importance in the dawning age of gravitational-wave astronomy. The development of novel approaches—capable to provide a consistent description of a variety of equilibrium and non-equilibrium properties, and based on dynamical models applicable over the whole relevant density range—will be needed to fully exploit the potential of future detections of gravitational wave signals.

This book has greatly benefited from countless discussions with our colleagues, collaborators, and students, whose advice and constructive criticisms we have deeply appreciated. Particular mention is owed to Ingo Sick, author of the picture appearing on the book cover,

Alessandro Lovato, who also contributed several figures, Kevin Schmidt, Francesco Pederiva, Sergio Rosati, and Artur Polls.

Finally, we would like to acknowledge how much we are indebted to the work of our late friends and collaborators Adelchi Fabrocini and Vijay R. Pandharipande, who gave fundamental and lasting contributions to the development of nuclear matter theory.

Throughout the book, we use a natural system of units, in which $\hbar = c = 1$, and the symbol ϱ denotes both the nucleon density, that is, the number of nucleons per fm^3, and the matter density, generally expressed in units of g cm^{-3}.

Omar Benhar and Stefano Fantoni. October, 2019

CHAPTER 1

INTRODUCTION

The concept of nuclear matter, which naturally emerges from the systematic analysis of observed nuclear properties, provides the foundation for the development of a unified theoretical framework, that can be used to model both atomic nuclei and the interior of neutron stars. In this chapter, we briefly outline the basis of the liquid drop model, according to which the nucleus can be described as an incompressible fluid, as well as the extension of this treatment to neutron star matter, whose structure and dynamics will be further discussed in Chapter 6.

1.1 NUCLEAR MATTER IN ATOMIC NUCLEI

The liquid drop model of the nucleus, first proposed by G. Gamow in 1932 [1], is based on a large body of data, providing information on nuclear binding energies and charge distributions .

The binding energy per nucleon in a nucleus of mass number A and charge Z is defined as

$$\frac{B(\mathrm{A},\mathrm{Z})}{\mathrm{A}} = \frac{1}{\mathrm{A}}\left[\mathrm{Z}m_p + \mathrm{N}m_n - M(\mathrm{A},\mathrm{Z})\right] , \tag{1.1}$$

where $\mathrm{N} = \mathrm{A} - \mathrm{Z}$ is the number of neutrons, while m_p, m_n and M denote the *measured* proton, neutron and nuclear mass, respectively. Note that, from the above definition, it follows that $B(\mathrm{Z},\mathrm{A})$ is a positive quantity.

Figure 1.1, displaying the A-dependence of $B(\mathrm{Z},\mathrm{A})/\mathrm{A}$ for all stable nuclei, shows that for $\mathrm{A} \gtrsim 20$ the binding energy becomes almost constant, its value being ~ 8.5 MeV. This observation indicates that the number of particles involved in nuclear interactions does not grow indefinitely with A, which in turn implies that nuclear forces have finite range.

Nuclear charge-density distributions are obtained from measurements of the electron-nucleus cross sections in the elastic scattering regime, in which the target nucleus is left in its ground state[1]. The observation that these distributions are nearly constant in the nuclear interior—their value, $\varrho_0 \approx 0.16\ \mathrm{fm}^{-3}$, being largely independent of A for $\mathrm{A} \gtrsim 16$—indicates that nuclei are nearly incompressible, that is, that nuclear forces become strongly repulsive at short distances. This feature, referred to as saturation of nuclear densities, is illustrated in Fig. 1.2, showing the radial dependence of the charge-densities of nuclei ranging from oxygen ($\mathrm{A} = 16$) to lead ($\mathrm{A} = 208$). The analysis of the available data also shows that the nuclear radius is simply related to the mass number, A, through $R_A = R_0\mathrm{A}^{1/3}$, with $R_0 \approx 1.2$ fm.

[1]Electron scattering experiments actually measure charge form factors, whose relation to the charge-density distribution is frame dependent. The results shown in this volume have been obtained in the so-called Breit frame.

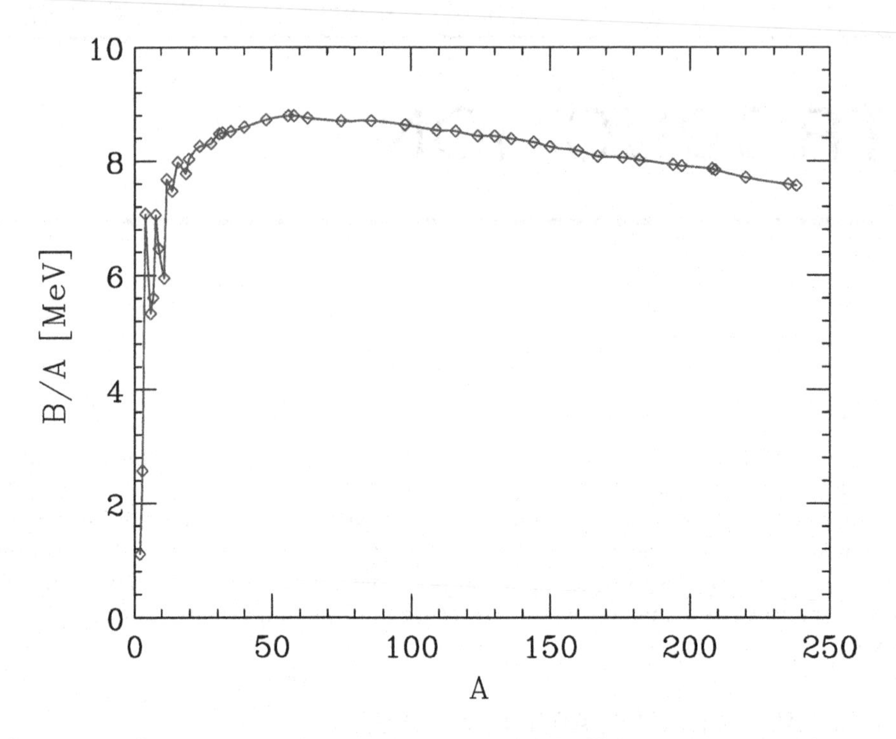

Figure 1.1 Mass dependence of the nuclear binding energy of stable nuclei, defined as in Eq. (1.1).

The main features emerging from the systematics of the nuclear binding energies and charge-density distributions point to a remarkable similarity between nuclear forces and the intermolecular forces determining the structure of a liquid drop, whose energy can be written in the form[2]

$$\mathcal{E} = -\alpha\mathcal{N} + \tau 4\pi\mathcal{R}^2 \ , \tag{1.2}$$

with $\mathcal{N}$, $\mathcal{R} \propto \mathcal{N}^{1/3}$ and τ being the number of molecules, the drop radius, and the surface tension of the liquid, respectively. From the above equation, it folllows that the binding energy per molecule depends on $\mathcal{N}$ according to

$$\frac{\mathcal{B}}{\mathcal{N}} = -\frac{\mathcal{E}}{\mathcal{N}} = \alpha - \beta\frac{1}{\mathcal{N}^{1/3}} . \tag{1.3}$$

The liquid drop analogy suggests that an expression similar to Eq. (1.3) may be used to describe the nuclear binding energy. Unlike molecules, however, protons are charged particles, and Coulomb forces must be taken into account.

Recalling that the potential energy associated with a charge Q confined to a spherical volume of radius R is $\propto Q^2/R$, and that Coulomb interactions between protons are repulsive, their contribution to the binding energy can be written in the form

$$B_{\rm C} = -\gamma\frac{\rm Z^2}{\rm A^{1/3}} \ , \tag{1.4}$$

[2]Throughout this volume, we will denote by $\mathcal{N}$ the number of particles in a generic many-body system, e.g. a molecule, and by A the number of nucleons in a nucleus, or, by extension, in nuclear matter.

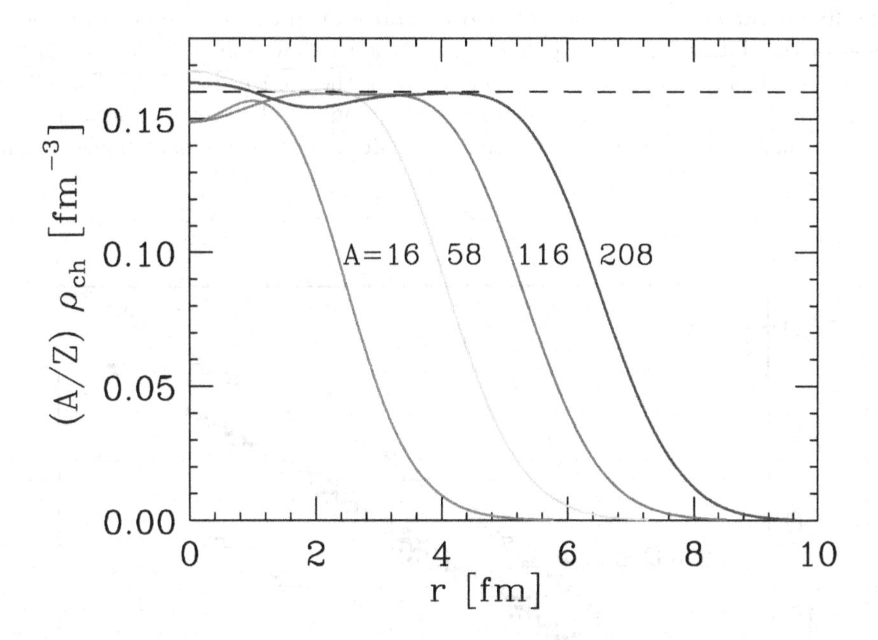

Figure 1.2 Radial dependence of the charge-density distributions of nuclei, normalised to the nuclear mass number A. The dashed horizontal line corresponds to the value $\varrho_0 = 0.16$ fm^{-3}.

with γ a positive constant.

Nuclear systematics suggests the inclusion of additional contributions to $B(\mathrm{A},\mathrm{Z})$. The observation that stable nuclei tend to have equal number of protons and neutrons, with the neutron excess $(\mathrm{N}-\mathrm{Z})/\mathrm{A}$ not exceeding ~ 0.2, can be explained considering that, as dictated by Pauli's exclusion principle, in nuclei with $\mathrm{Z} = \mathrm{N}$ protons and neutrons occupy the lowest $\mathrm{A}/2$ energy levels, and turning a proton into a neutron requires an energy $\Delta \sim \mathrm{A}^{-1}$. This feature can be described adding to $B(\mathrm{A},\mathrm{Z})$ a *symmetry* term

$$B_{\mathrm{S}} = -\delta \frac{(\mathrm{N}-\mathrm{Z})^2}{\mathrm{A}} = -\delta \frac{(\mathrm{A}-2\mathrm{Z})^2}{\mathrm{A}} \ , \tag{1.5}$$

with $\delta > 0$.

Finally, the empirical evidence that nuclei with even numbers of protons and neutrons are energetically favoured can be accounted for with the inclusion of a term

$$B_{\mathrm{P}} = -\epsilon \frac{1}{\mathrm{A}^{1/2}} \ , \tag{1.6}$$

where $\epsilon = \pm\epsilon_0$ and $\epsilon_0 > 0$. The plus and minus signs apply, respectively, to the cases of odd or even values of both N and Z, corresponding to even A, while for odd A $\epsilon = 0$.

Collecting all the above contributions, the nuclear binding energy per nucleon can be cast in the form,

$$\frac{B(\mathrm{A},\mathrm{Z})}{\mathrm{A}} = \alpha - \beta \frac{1}{\mathrm{A}^{1/3}} - \gamma \frac{\mathrm{Z}^2}{\mathrm{A}^{4/3}} - \delta \frac{(\mathrm{A}-2\mathrm{Z})^2}{\mathrm{A}^2} - \epsilon \frac{1}{\mathrm{A}^{3/2}}, \tag{1.7}$$

known as von Weitzäker semi empirical mass formula [2]. By properly adjusting the values of the five parameters involved, the above expression provides a remarkably accurate description of the data shown in Fig. 1.1, except for the few points corresponding to the spikes at low A. The liquid drop model largely explains the observed pattern of stable nuclei, illustrated in Fig. 1.3, and provides accurate predictions of the energy released in fission reactions, in which a nucleus of mass number A breaks down into two nuclei of smaller masses.

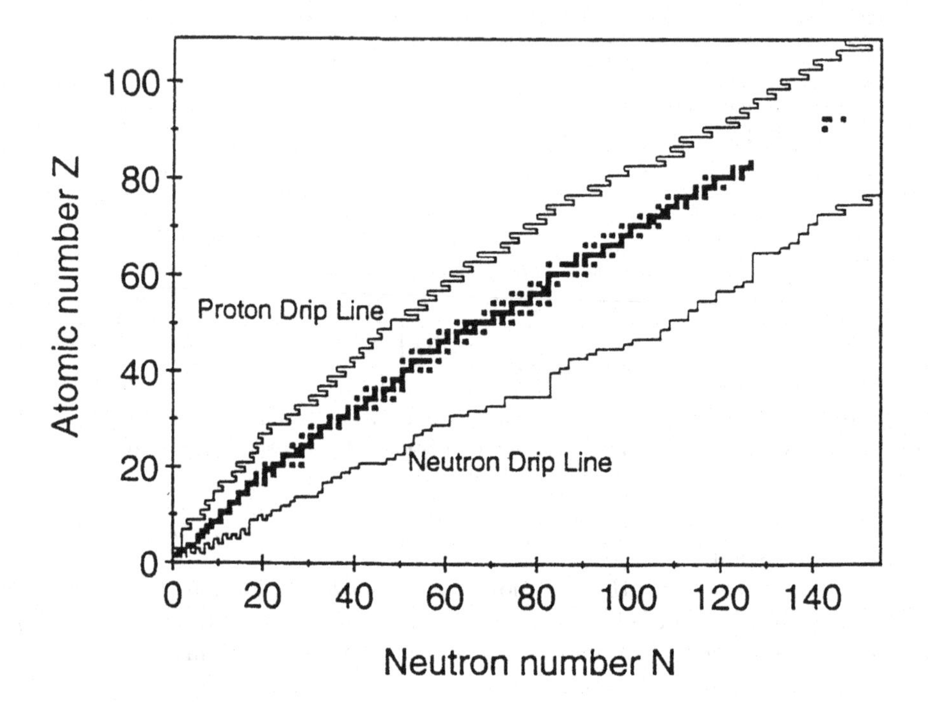

Figure 1.3 Chart of the nuclides. The black squares represent stable nuclei as a function of their charge Z and neutron number N = A − Z.

In the A → ∞ limit, Eq. (1.7) can be employed to obtain the binding energy per nucleon of nuclear matter, defined as a uniform system consisting of infinite numbers of protons and neutrons subject to strong interactions only. In the case of isospin-symmetric matter, with N = Z, one finds

$$\frac{B(\mathrm{A},\mathrm{Z})}{\mathrm{A}} = -\frac{E_0}{\mathrm{A}} = \alpha \approx 16\ \frac{\mathrm{MeV}}{\mathrm{A}}\ , \tag{1.8}$$

where E_0 can be identified as the ground-state energy of the system. The corresponding density, ϱ_0, can be inferred from the charge-density distributions shown in Fig. 1.2, suggesting that

$$\varrho_0 = \lim_{\mathrm{A}\to\infty} \frac{\mathrm{A}}{\mathrm{Z}}\ \varrho_{\mathrm{ch}}(r=0) = 0.16\ \mathrm{fm}^{-3}\ , \tag{1.9}$$

with A/Z = 2.

1.2 NUCLEAR MATTER IN NEUTRON STARS

Figure 1.3 shows that, owing to their repulsive electrostatic interactions, the number of protons in stable nuclei is limited to $\sim$ 80. For large neutron excess, on the other hand, nuclei become unstable against β-decays turning neutrons into protons.

The existence of compact astrophysical objects made of neutrons, the stability of which is the result of gravitational attraction, was first proposed by L. Landau in 1932, shortly after the discovery of the neutron [3]. In 1934, W. Baade and F. Zwicky suggested that a neutron star may be formed in the aftermath of a supernova explosion [4]. Finally, in 1968 the newly observed pulsars, radio sources emitting pulses at a constant frequency, were identified with highly magnetised rotating neutron stars [5].

The results of a pioneering study, carried out in 1939 by J. Oppenheimer and G. Volkoff within the framework of general relativity [6], show that the mass of a star consisting of non interacting neutrons cannot exceed $\sim 0.8\ M_\odot$, where $M_\odot = 1.989 \times 10^{30}$ kg denotes the solar mass. The inconsistency between this value and the observed neutron star masses, typically $M_{\rm NS} \sim 1.4\ M_\odot$, demonstrates that in these systems hydrostatic equilibrium requires a pressure other than the degeneracy pressure predicted by Fermi-Dirac statistics, the origin of which has to be traced back to the occurrence of interactions between the constituent neutrons.

Combining the measured neutron star masses with the available experimental information on their radius, pointing to values $R_{\rm NS} \sim 10$ km, one obtains an average density, $\varrho_{\rm NS} \gtrsim 10^{14}$ g/cm^{-3}, comparable to the central density of atomic nuclei, ϱ_0[3]. A straightforward order-of-magnitude calculation of the number of constituent neutrons yields $N_{\rm NS} \sim 10^{57}$.

The internal structure of a neutron star, schematically represented in Fig. 1.4, is believed to feature a sequence of layers of different composition. While the properties of matter in the outer crust—corresponding to densities ranging from $\sim 10^7$ g/cm^3 to the so-called neutron drip density, $\varrho_{\rm drip} \sim 4 \times 10^{11}$ g/cm^3—can be inferred from nuclear data, models of matter at $\varrho_{\rm drip} < \varrho < 2 \times 10^{14}$ g/cm^3 are largely based on extrapolations of the available empirical information, as the extremely neutron rich nuclei appearing in this density regime are not observed on earth.

The density of the neutron star core ranges between $\sim \varrho_0$, at the boundary with the inner crust, and a central value that can be as large as $1 - 4 \times 10^{15}$ g/cm^3, depending on the star mass and on the properties of matter in its interior. All models based on hadronic degrees of freedom predict that in the density range $\varrho_0 \lesssim \varrho \lesssim 2\varrho_0$ neutron star matter consists mainly of neutrons, with the admixture of a small number of protons, electrons and muons. At any given density, the fraction of protons and leptons is determined by the requirements of weak equilibrium and charge neutrality. Most calculations suggest that this fraction is rather small, of the order of $\sim$10% at most. Hence, for many applications, modelling neutron star matter with pure neutron matter can be regarded as a reasonable approximation.

The picture may change significantly at larger density, with the appearance of heavier strange baryons produced in weak interaction processes. For example, although the mass of the Σ^- exceeds the neutron mass by more than 250 MeV, the reaction $n + e^- \to \Sigma^- + \nu_e$ becomes energetically allowed as soon as the sum of the neutron and electron chemical potentials becomes equal to the Σ^- chemical potential.

Finally, as nucleons are known to be composite objects of size in the range $\sim 0.5 - 1.0$ fm,

[3]The equilibrium density of isospin-symmetric nuclear matter, $\varrho_0 = 0.16$ fm^{-3}, corresponds to a matter density $\rho_0 = 2.67 \times 10^{14}$ g/cm^3.

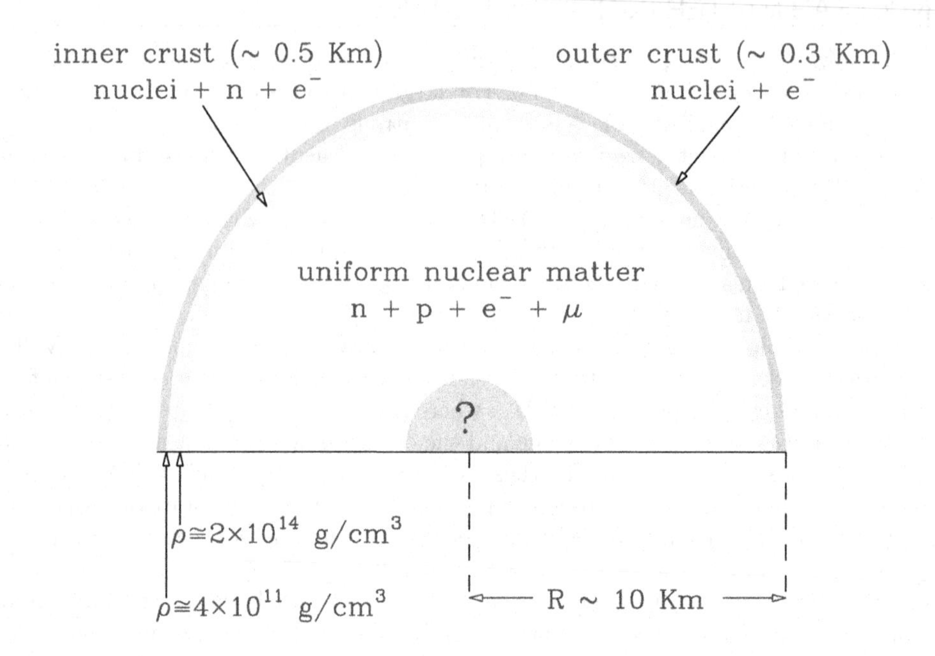

Figure 1.4 Schematic representation of internal structure of a neutron star.

corresponding to a density $\sim 10^{15}$ g/cm^3, it is expected that, if the density of the neutron star core reaches this value, matter undergoes a transition to a new phase, predicted by the fundamental theory of strong interactions, in which quarks are no longer clustered into nucleons or hadrons.

CHAPTER 2

NUCLEAR DYNAMICS

While Quantum Chromo-Dynamics, or QCD, has been long recognised as the fundamental theory of strong interactions, its applications are largely limited to the high-energy regime, in which the elementary degrees of freedom of the theory manifest themselves, and their interactions can be treated in perturbation theory. At lower energy, on the other hand, QCD becomes non perturbative, and the fundamental approach is based on lattice calculations involving non trivial difficulties.

In this chapter, we review a more phenomenological approach, in which nucleons are treated as effective degrees of freedom, whose interactions are described within models constrained by the available data.

2.1 THE PARADIGM OF MANY-BODY THEORY

The observations of nuclear properties indicate that, to a remarkably large extent, atomic nuclei—and, by extension, nuclear matter—can be described as non relativistic systems consisting of point-like particles, the dynamics of which are dictated by a Hamiltonian of the form

$$H = \sum_{i=1}^{A} \frac{\mathbf{p}_i^2}{2m} + \sum_{j>i=1}^{A} v_{ij} + \sum_{k>j>i=1}^{A} V_{ijk} \ . \tag{2.1}$$

In the above equation, $\mathbf{p}_i$ and m denote the momentum of the i-th nucleon and its mass, while the potentials v_{ij} and V_{ijk} account for two- and three-nucleon interactions, respectively. Note that the potentials have a non trivial operator structure, involving a dependence on the discrete quantum numbers specifying the state of the interacting particles.

Before analysing the main features of the potentials appearing in Eq.(2.1), a critical discussion of the tenet underlying the paradigm of nuclear many-body theory is in order.

Clearly, the assumption that protons and neutrons can be described as point-like particles needs to be reconciled with the observation that nucleons have in fact finite size. The nucleon radius can be inferred from the proton charge distribution, measured by elastic electron scattering on hydrogen. The value resulting from a state-of-the-art analysis of the data turns out to be $\langle r_{ch}^2 \rangle^{1/2} = 0.887 \pm 0.012$ fm [7].

Figure 2.1 shows the charge-density profiles of two protons separated by a distance d = 1.6 fm (upper panel) and 1.0 fm (lower panel), computed using the parametrisation of the measured proton from factors of Bradford *et al* [8]. It is apparent that at $d = 1.6$ fm—the average nucleon-nucleon (NN) separation distance in nuclei such as carbon or oxygen—the overlap is marginal, and the point-like approximation is expected to be applicable. On the other hand, the lower panel suggests that at shorter separation distance, corresponding to higher nucleon density, the description based on nuclear many-body theory may become inadequate.

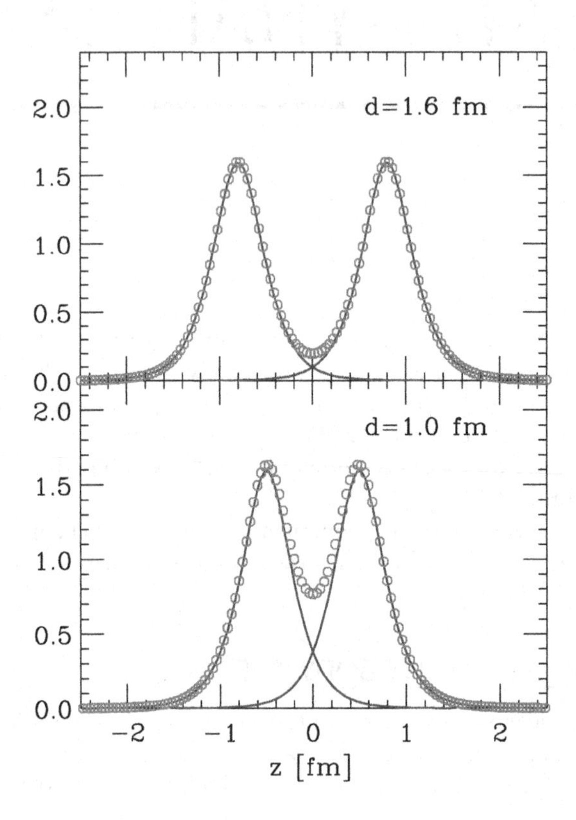

Figure 2.1 The solid lines show the charge-density distributions of two protons separated by a distance $d = 1.6$ fm (upper panel) and 1.0 fm (lower panel). The sum of the two distribution is represented by the diamonds.

The validity of the description in terms of nucleons down to distances of the order of 1 fm is also supported by electron-nucleus scattering data. The observation of y-scaling in experiments performed using a variety of targets, ranging from ^{2}H to nuclei as heavy as ^{197}Au, unambiguously shows that at momentum transfer $\gtrsim 1$ GeV and negative y the beam particles couple to nucleons, carrying momenta up to ~ 700 MeV [9].

2.2 EMPIRICAL FACTS ON NUCLEAR FORCES

Some of the prominent features of the NN interaction can be deduced from the analysis of nuclear systematics. They can be summarised as follows.

- The saturation of nuclear density, discussed in Section 1.1, suggests that the NN potential is strongly repulsive at short distances, i.e. that

$$v(r_{ij}) > 0 \quad , \quad r_{ij} < r_c \; , \tag{2.2}$$

  where r_c denotes the radius of the repulsive core.

- The observation that the binding energy per nucleon of a nucleus of mass number A and charge Z, defined as in Eq.(1.1), is nearly constant for all nuclei with A $\geq$ 20 indicates that nuclear forces have finite range r_0, i.e. that

$$v(r_{ij}) = 0 \quad , \quad r_{ij} > r_0 \ , \tag{2.3}$$

with $r_0 \ll R_A$.

- The spectra of the so called mirror nuclei, i.e. pairs of nuclei having the same mass number A and charges differing by one unit, exhibit striking similarities[1]. The observation that the energies of the levels with the same parity and angular momentum are the same, up to small electromagnetic corrections, suggests that protons and neutrons have similar nuclear interactions, i.e that nuclear forces are charge symmetric.

Charge symmetry is the manifestation of a more general property of nuclear forces, referred to as isotopic invariance. Neglecting the $\sim 0.1\%$ mass difference, proton and neutron can be viewed as two states of the same particle, the nucleon (N), labeled by a quantum number dubbed isospin.

A nucleon in vacuum can be described by the Dirac equation obtained from the Lagrangian density

$$\mathcal{L} = \bar{\psi}_N \left(i\partial\!\!\!/ - m \right) \psi_N \ , \tag{2.4}$$

where

$$\psi_N = \left(\begin{array}{c} \phi_p \\ \phi_n \end{array} \right) \ , \tag{2.5}$$

ϕ_p and ϕ_n being the four-component spinors associated with the proton and the neutron, respectively. In Eq. (2.4), $\partial\!\!\!/ = \gamma_\mu \partial^\mu$, where γ_μ denotes a Dirac gamma matrix, and $m \approx 939$ MeV is the nucleon mass.

The Lagrangian density (2.4) is invariant under the SU(2) global phase transformation

$$U = \mathrm{e}^{i\alpha_j \tau_j} \ , \tag{2.6}$$

where the α_j ($j = 1, 2, 3$) are constants, independent of the coordinate x, and the τ_j are Pauli matrices acting in isospin space.

The above equations show that the nucleon is described by a isospin doublet, with proton and neutron corresponding to isospin projections $+1/2$ and $-1/2$, respectively. Proton-proton and neutron-neutron pairs always have total isospin $T = 1$, whereas a proton-neutron pair may have either $T = 0$ or $T = 1$. The two-nucleon isospin states $|T, T_3\rangle$ can be specified as follows

$$|1, 1\rangle = |pp\rangle \quad , \quad |1, 0\rangle = \frac{1}{\sqrt{2}} \left(|pn\rangle + |np\rangle \right) \quad , \quad |1, -1\rangle = |nn\rangle \ ,$$

$$|0, 0\rangle = \frac{1}{\sqrt{2}} \left(|pn\rangle - |np\rangle \right) \ .$$

Isospin invariance implies that the interaction between two nucleons separated by a distance r and having total spin S depends on their total isospin T, but not on the projection T_3. For example, the potential $v(r)$ acting between two protons, or two neutrons, with spins coupled to $S = 0$ is the same as the potential acting between a proton and a neutron with spins and isospins coupled to $S = 0$ and $T = 1$.

[1]The number of protons in a nucleus belonging to a mirror pair is the same as the number of neutrons in its companion. For example, $^{15}_{7}$N ($A = 15$, $Z = 7$) and $^{15}_{8}$O ($A = 15$, $Z = 8$) are mirror nuclei.

2.3 PHENOMENOLOGICAL POTENTIALS

In this section, we will briefly describe the derivation of the potentials v_{ij} and V_{ijk} based on the analysis of the properties of the two- and three-nucleon systems, which can be obtained from exact calculations.

2.3.1 The nucleon-nucleon potential

The details of NN forces are best analysed in the two-nucleon system. There is only one NN bound state, the nucleus of deuterium, or deuteron (^{2}H), consisting of a proton and a neutron with total spin and isospin $S = 1$ and $T = 0$. Note that this is in itself a clear manifestation of the strong spin depencence of NN interactions.

Another important piece of information is obtained from the observation that the deuteron has a non vanishing electric quadrupole moment, reflecting a non spherically symmetric charge distribution. This is an unambiguous indication that the NN forces are non central.

In addition to the properties of the two-nucleon bound state, the large data base of phase shifts measured in proton-proton and proton-neutron scattering experiments provides valuable complementary information, which has been extensively exploited to test and constrain models of nuclear dynamics.

A theoretical description of NN interactions based on the formalism of quantum field theory was first proposed by H. Yukawa in 1935 [10]. In his seminal paper, Yukawa made the hypothesis that nucleons interact through the exchange of a particle whose mass, μ, can be deduced from the interaction range, r_0, exploiting the relation

$$r_0 \sim \frac{1}{\mu} \ . \tag{2.7}$$

For $r_0 \sim 1$ fm, one finds $\mu \sim 200$ MeV (1 fm^{-1} = 197.3 MeV).

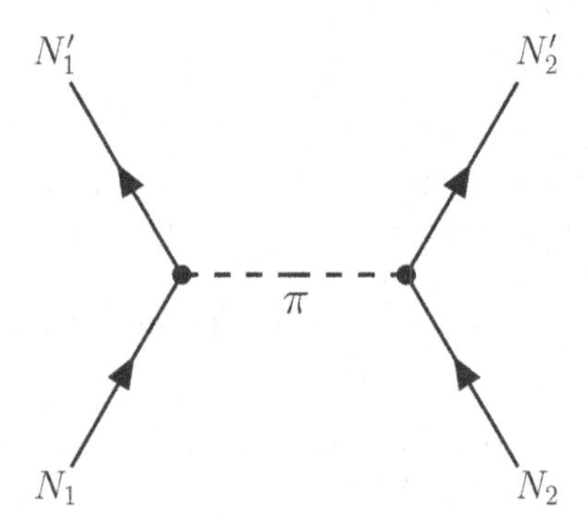

Figure 2.2 Feynman diagram describing the one-pion-exchange process in NN scattering. The corresponding amplitude is given by Eq. (2.8).

Yukawa's suggestion was successfully implemented identifying the exchanged particle with the π-meson, or pion, with mass $m_\pi \sim 140$ MeV. Experiments have shown that the pion has spin-parity 0^-, and three charge states, denoted π^+, π^0 and π^-. Hence, it can be regarded as an isospin $T = 1$ triplet, the charge states being associated with isospin projections $T_3 = +1, 0$ and -1, respectively[2].

The simplest π-nucleon coupling compatible with the observation that nuclear interactions conserve parity has the pseudoscalar form $ig\gamma^5\boldsymbol{\tau}$, where $\gamma^5 = i\gamma^0\gamma^1\gamma^2\gamma^3$, g is the coupling constant and the operator $\boldsymbol{\tau}$ describes the isospin of the nucleon. With this choice of the interaction vertex, the invariant amplitude of the process depicted in Fig. 2.2 can be readily written, using standard Feynman's diagram techniques, as

$$\mathcal{M} = -ig^2 \, \frac{\bar{u}(p_2', s_2')\gamma_5 u(p_2, s_2)\bar{u}(p_1', s_1')\gamma_5 u(p_1, s_1)}{k^2 - m_\pi^2} \langle \boldsymbol{\tau}_1 \cdot \boldsymbol{\tau}_2 \rangle \, , \tag{2.8}$$

where $k = p_1' - p_1 = p_2 - p_2'$, $k^2 = k_\mu k^\mu = k_0^2 - \mathbf{k}^2$, and $u(p, s)$ is the Dirac spinor associated with a nucleon of four-momentum $p \equiv (E, \mathbf{p})$, with E$=\sqrt{\mathbf{p}^2 + m^2}$, and spin projection s. Finally

$$\langle \boldsymbol{\tau}_1 \cdot \boldsymbol{\tau}_2 \rangle = (\eta_2'^\dagger \boldsymbol{\tau} \eta_2) \cdot (\eta_1'^\dagger \boldsymbol{\tau} \eta_1) \, , \tag{2.9}$$

η_i being the two-component Pauli spinor describing the isospin state of nucleon i.

In the non relativistic limit, Yukawa's theory leads to define the one-pion-exchange (OPE) potential, that can be written in coordinate space in the form

$$\begin{aligned} v_\pi &= \frac{g^2}{4} \frac{m_\pi}{m^2} \, (\boldsymbol{\tau}_1 \cdot \boldsymbol{\tau}_2)(\boldsymbol{\sigma}_1 \cdot \boldsymbol{\nabla})(\boldsymbol{\sigma}_2 \cdot \boldsymbol{\nabla}) \, \frac{e^{-x}}{x} \\ &= \frac{g^2}{(4\pi)^2} \, \frac{m_\pi^3}{4m^2} \, \frac{1}{3} (\boldsymbol{\tau}_1 \cdot \boldsymbol{\tau}_2) \left\{ \left[(\boldsymbol{\sigma}_1 \cdot \boldsymbol{\sigma}_2) + S_{12} \left(1 + \frac{3}{x} + \frac{3}{x^2} \right) \right] \frac{e^{-x}}{x} \right. \\ &\quad \left. - \, \frac{4\pi}{m_\pi^3} (\boldsymbol{\sigma}_1 \cdot \boldsymbol{\sigma}_2) \delta^{(3)}(\mathbf{r}) \right\} \, , \end{aligned} \tag{2.10}$$

where $\delta^{(3)}$ denotes the three-dimensional δ-function, $x = m_\pi r$, the matrices $\boldsymbol{\sigma}_i$ act in spin space and the tensor operator

$$S_{12} = \frac{3}{r^2} (\boldsymbol{\sigma}_1 \cdot \mathbf{r})(\boldsymbol{\sigma}_2 \cdot \mathbf{r}) - (\boldsymbol{\sigma}_1 \cdot \boldsymbol{\sigma}_2) \, , \tag{2.11}$$

is reminiscent of the one describing the non-central interaction between two magnetic dipoles.

For $g^2/(4\pi) \approx 14$, the OPE potential provides a fairly good description of the long range ($r \gtrsim 1.5$ fm) component of the NN interaction, as shown by the analysis of the phase shifts corresponding to scattering in states of high angular momentum. Owing to the presence of a strong centrifugal barrier, in these states the probability of finding the two interacting nucleons at small relative distance is in fact strongly suppressed.

At intermediate and short range, Eq.(2.7) suggests that more complicated processes, involving the exchange of two pions or heavier particles—such as the ρ and ω mesons , whose masses are $m_\rho = 770$ MeV and $m_\omega = 782$ MeV, respectively—must be taken into consideration. Moreover, when the relative distance becomes very small—typically $r \lesssim 0.5$ fm—nucleons, being composite and finite in size, are expected to overlap, as illustrated in Fig. 2.1. In this regime, NN forces should in principle be described in terms of interactions involving the nucleon constituents—quarks and gluons—as dictated by the fundamental theory of strong interactions.

[2]The pion spin has been deduced from the balance of the reaction $\pi^+ + {}^2\text{H} \leftrightarrow p + p$. The intrinsic parity was determined observing π^- capture from the K shell of the deuterium atom, leading to the appearance of two neutrons, $\pi^- + d \to n + n$.

Studies aimed at deriving the NN potential from lattice QCD calculations have recently achieved remarkable progress in predicting its qualitative features. As an example, Fig. 2.3 reports a comparison between the potentials obtained by Ishii *et al.* [11] and the predictions of the OPE model. However, the results of nuclear matter calculations performed using a lattice QCD potential suggest that significant developments will be needed to reach the level required to explain the empirical data in a fully quantitative fashion [12].

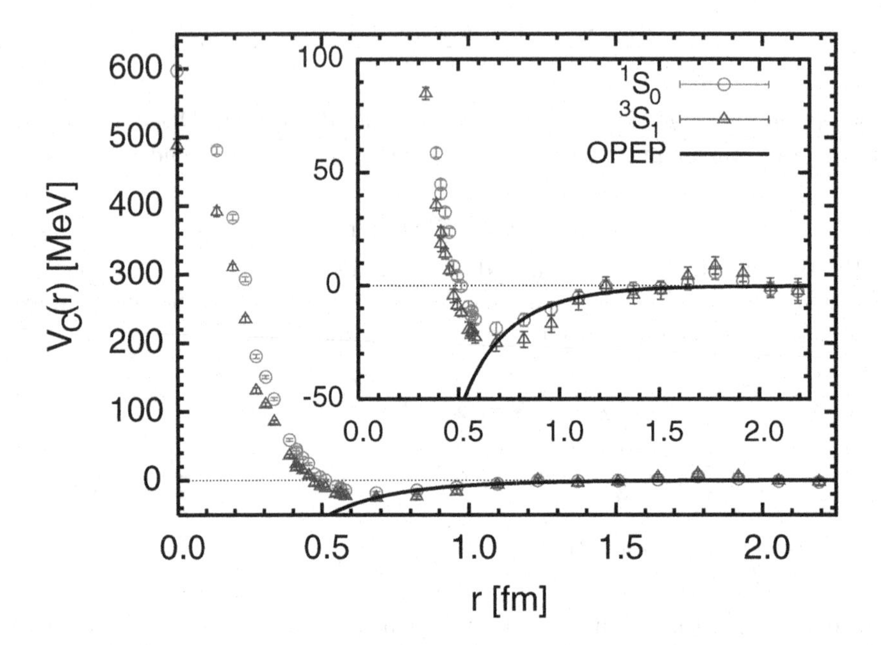

Figure 2.3 Radial dependence of the NN potentials in the singlet and triplet S channels, obtained by Ishii *et al.* [11] from lattice calculations at pion mass $m_\pi = m_\rho/2$, $m_\rho = 770$ MeV being the ρ-meson mass. For comparison, the solid lines show the long-range behaviour of the one-pion-exchange potential (OPEP).

Phenomenological potentials describing the NN interaction at all ranges can be conveniently written in the form

$$v_{ij} = \sum_p v^p(r_{ij}) O^p_{ij} \ , \tag{2.12}$$

where the functions v^p only depend on the distance between the interacting particles, $r_{ij} = |\mathbf{r}_i - \mathbf{r}_j|$, while the operators O^p_{ij} account for the spin-isospin dependence of NN interactions, as well as for the presence of non-central forces. The most important contributions to the sum appearing in the right-hand side of Eq.(2.12) are those associated with the operators

$$O^{p\leq 6}_{ij} = [1, (\boldsymbol{\sigma}_i \cdot \boldsymbol{\sigma}_j), S_{ij}] \otimes [1, (\boldsymbol{\tau}_i \cdot \boldsymbol{\tau}_j)] \ , \tag{2.13}$$

with S_{12} given by Eq.(2.11). Note that the OPE potential of Eq.(2.10) can also be written in terms of the six operators of Eq.(2.13).

State-of-the-art phenomenological models of v_{ij}, such as the Argonne v_{18} (AV18) potential [13]—determined from an accurate fit of the NN scattering phase shifts up to pion production threshold, the low-energy NN scattering parameters and deuteron properties—include twelve additional terms. The operators corresponding to $p = 7, \ldots, 14$ are associated with the non-static components of the NN interaction, while those corresponding to $p = 15, \ldots, 18$ take into account small violations of charge symmetry. The full AV18 potential involves 40 adjustable parameters, and fits the 4301 phase shifts collected in the Nijmegen data base with a reduced χ-square of 1.09.

A somewhat simplified interaction, referred to as Argonne v_6' (AV6P) , has been constructed projecting the full AV18 onto the basis of the six operators of Eq. (2.13) [14]. This potential, designed for easier use in many-body calculations, reproduces the deuteron binding energy and electric quadrupole moment with accuracy of 1% and 5%, respectively, and provides an excellent fit of the phase shifts in the $^1S_0$ channel, corresponding to $T = 1$, $S = 0$ and angular momentum $\ell = 0$.

The energy dependence of the $^1S_0$ phase shifts is illustrated in Fig. 2.4. It is apparent that the results obtained using the AV6P and AV18 potentials are nearly indistinguishable from one another, and provide an accurate description of the data resulting from the analyses of the Nijmengen group [15, 16] and Workman *et al.* [17] up to beam energies $\sim$ 600 MeV, well beyond the pion production threshold, $E_{\text{thr}} \approx 350$ MeV.

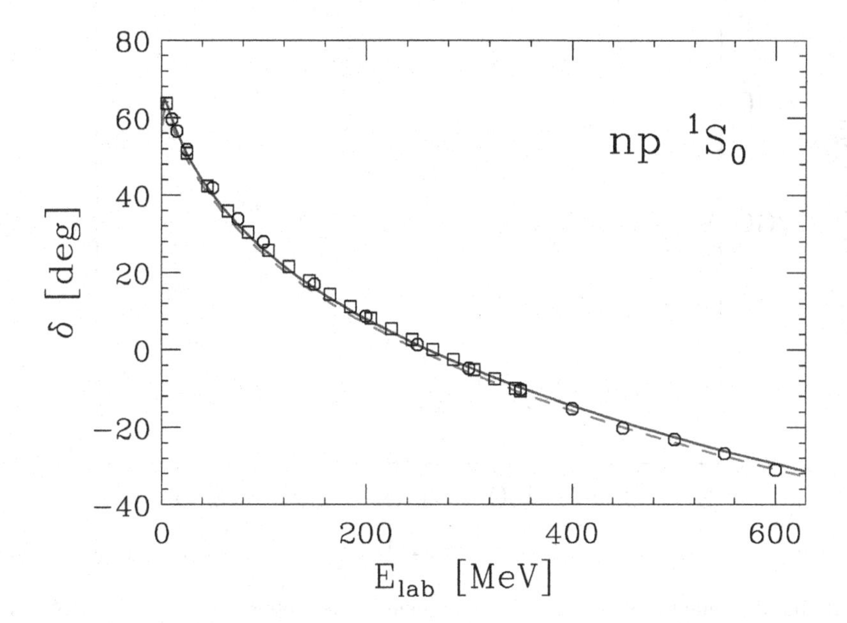

Figure 2.4 Neutron-proton scattering phase shitfs in the $^1S_0$ channel as a function of beam energy in the lab frame. The solid and dashed lines correspond to results obtained using the AV18 and AV6P potentials, respectively. Squares and circles represent data from the analyses of the NIjmegen Group [15, 16] and Workman *et al.* [17].

Projection of the AV18 potential onto a basis of eight operators, allowing to include the effects of spin-orbit interactions in both the singlet and triplet isospin channels, provides an improved description of the P-states phase shifts. The resulting interaction is dubbed Argonne v_8' (AV8P) potential [14].

It has to be emphasised that the ability to explain the data at large energy is critical to the application of a potential model to describe the properties of nuclear matter in the high-density region, relevant to neutron star physics. To see this, consider a scattering process involving two nucleons embedded in the nuclear medium at density ϱ. In the nearly degenerate regime typical of neutron stars, the collisions only involve particles with energies close to the Fermi energy, $e_F \propto \varrho^{2/3}$. As a consequence, a simple relation can be established between the energy of the projectile particle in the laboratory frame and the matter density. In the case of head-on scattering in pure neutron matter one finds

$$E_{\rm lab} = \frac{1}{m}(3\pi^2 \varrho)^{2/3} , \tag{2.14}$$

implying that the maximum energy of Fig. 2.4 corresponds to $\varrho \approx 4\varrho_0$, where $\varrho_0 = 0.16$ fm^{-3} is the central density of atomic nuclei, see Fig. 1.2.

The prominent features of the NN potential in the $^1S_0$ state are shown in Fig. 2.5. The short-range repulsive core, to be ascribed to heavy-meson exchange or to more complicated mechanisms involving nucleon constituents, is followed by an intermediate range attractive region, largely due to correlated two-pion exchange processes. Finally, at long range the one-pion-exchange mechanism dominates.

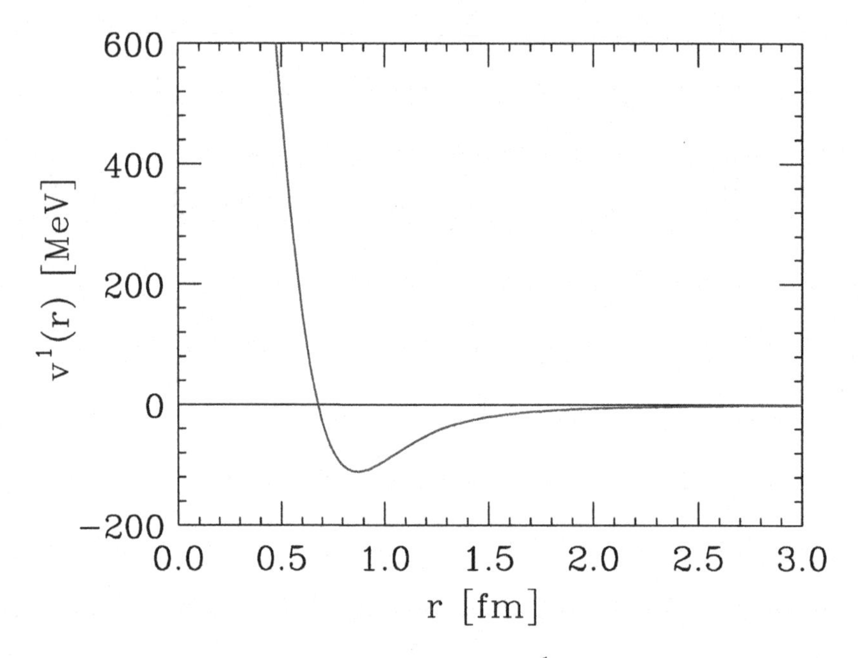

Figure 2.5 Radial dependence of the central component, v^1 component of the AV6P potential, see Eq.(2.12).

2.3.2 Three-nucleon forces

Phenomenological NN potentials reproduce the observed properties of the two-nucleon system by construction. However, they fail to explain the ground-state energy of the three-nucleon bound states, the nuclei ^{3}H and ^{3}He, whose values can be computed exactly using deterministic techniques. In order to bring theoretical results into agreement with the data, a three-nucleon (NNN) potential must be added to the nuclear Hamiltonian, as shown in Eq.(2.1).

The inclusion of irreducible three-body forces is long known to be needed to describe the interactions of composite systems without explicitly considering their internal structure. The three-body system comprising the Earth, the Moon and a satellite orbiting the Earth provides an archetypal example [18]. In this case, a three body-force is required to account for the tidal deformation of the Earth, which explicitly depends on the position of the moon and affects at the same time the orbital motion of the satellite.

The nature of nuclear three-body forces is clearly highlighted in the seminal paper of J. Fujita and H. Miyiazawa [19]. These authors argued that the most important mechanism at work is the two-pion-exchange process in which a NN interaction leads to the excitation of one of the participating nucleons to a Δ resonance, of mass $M_\Delta \approx 1232$ MeV, which then decays in the aftermath of the interaction with a third nucleon.

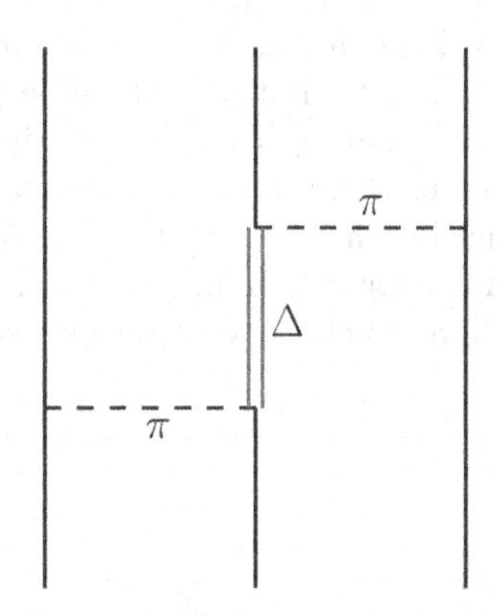

Figure 2.6 Diagrammatic representation of the Fujita-Miyazawa three-nucleon interaction, arising from two-pion exchange processes in which one of the nucleons is excited to a Δ resonance in the intermediate state.

Commonly used phenomenological models of the NNN force, such as the Urbana IX (UIX) potential [20], are written in the form

$$V_{ijk} = V_{ijk}^{2\pi} + V_{ijk}^{R} \ , \tag{2.15}$$

where $V_{ijk}^{2\pi}$ is the attractive Fujita-Miyazawa term, a schematic representation of which is given in Fig. 2.6, while V_{ijk}^{N} is a purely phenomenological repulsive term.

The explicit expressions of the two contributions appearing in the right-hand side of Eq.(2.15) are

$$V_{ijk}^{2\pi} = A_{2\pi} \sum_{\text{cycl}} \left\{ \{X_{ij}, X_{jk}\}\{\boldsymbol{\tau}_i \cdot \boldsymbol{\tau}_j, \boldsymbol{\tau}_j \cdot \boldsymbol{\tau}_k\} + \frac{1}{4}[X_{ij}, X_{jk}][\boldsymbol{\tau}_i \cdot \boldsymbol{\tau}_j, \boldsymbol{\tau}_j \cdot \boldsymbol{\tau}_k] \right\} \ , \tag{2.16}$$

and

$$V_{ijk}^{R} = U \sum_{\text{cycl}} T^2(m_\pi r_{ij}) T^2(m_\pi r_{jk}) \ , \tag{2.17}$$

where the sums are extended to all cyclic permutations of the indices i, j and k. In the above equations

$$X_{ij} = Y(m_\pi r_{ij}) \boldsymbol{\sigma}_i \cdot \boldsymbol{\sigma}_j + T(m_\pi r_{ij}) S_{ij} \ , \tag{2.18}$$

with

$$Y(x) = \frac{e^{-x}}{x}\xi(r) , \tag{2.19}$$

$$T(x) = (1 + \frac{3}{x} + \frac{3}{x^2})Y(x) , \tag{2.20}$$

where the function ξ cuts off the contribution of OPE interactions at short distances, and S_{ij} is given by Eq.(2.11). The strength $A_{2\pi}$ and U are determined in such a way as to reproduce the binding energies of ^{3}He and ^{4}He and the empirical equilibrium density of isospin-symmetric nuclear matter, hereafter referred to as SNM, respectively.

The nuclear Hamiltonians constructed supplementing the AV18 NN potential with a phenomenological NNN potentials, such as the UIX model or the more advanced model referred to as Illinois-7 (IL7) [21], while being mainly constrained to reproduce the properties of the two- and three-nucleon systems, exhibit a remarkable predictive power. The results of Quantum Monte Carlo (QMC) calculations carried out using the AV18-IL7 Hamiltonian, extensively reviewed by Carlson *et al.* [22], account for the measured energies of the ground and low-lying excited states of nuclei with $A \leq 12$ with accuracy of few percent, see Fig. 2.7.

Calculations based on phenomenological Hamiltonians also account for of variety electroweak nuclear observables of light nuclei, including electromagnetic form factors. and low-energy transition rates [23].

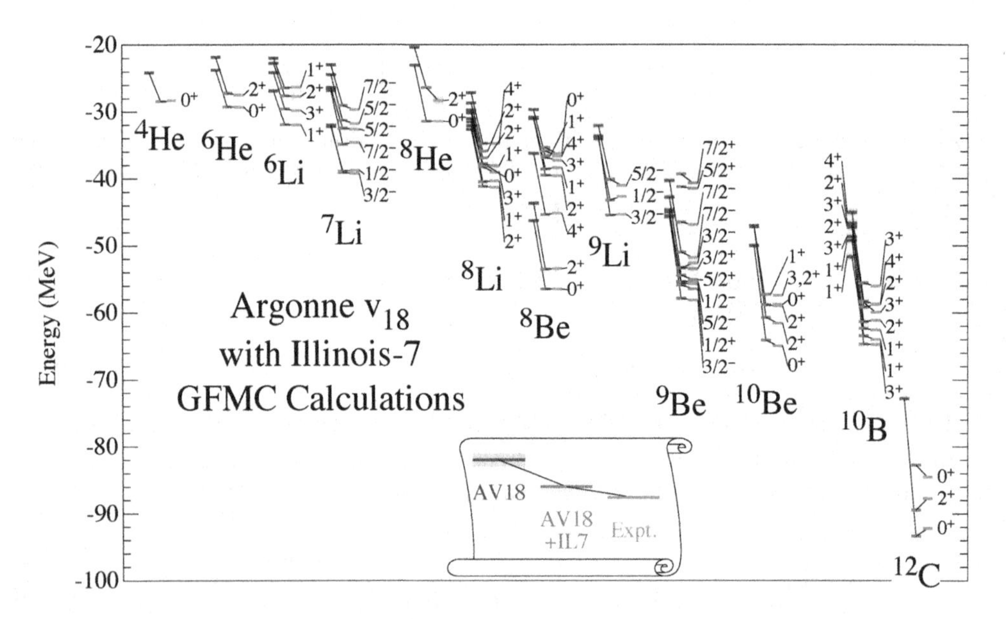

Figure 2.7 Comparison between the spectra of light nuclei obtained by Carlson *et al.* using the Green's Function Monte Carlo (GFMC) technique and experimenta data. The calculations have been carried out using a nuclear Hamiltonian comprising the AV18 NN potential, with and without inclusion of the IL7 model of NNN interactions.Taken from [22].

Three-nucleon interactions play a critical role in nuclear matter. Their inclusion is essential to explain the equilibrium properties of SNM, and strongly affects the equation of state—that is, the density dependence of the ground-state expectation value of the nuclear Hamiltonian—at densities $\varrho > \varrho_0$.

As mentioned above, most phenomenological three-nucleon potentials include a parameter adjusted in such a way as to obtain saturation in SNM at $\varrho \approx \varrho_0$ using the advanced many-body approaches to be discussed in Chapters 4 and 5. It is remarkable, however, that the occurrence of a minimum of the energy per nucleon, $E(\varrho)$, at $\varrho \lesssim \varrho_0$ is also predicted by an improved version of the Tucson-Melbourne phenomenological potential, not tuned to reproduce the properties of SNM [24].

It has to be pointed out, however, that, in spite of being essential to reproduce the properties of the few-nucleon systems, the contribution of the NNN potential to the ground-state expectation value of the nuclear Hamiltonian turns out to be quite small. Typically, one finds $\langle V_{ijk} \rangle / \langle v_{ij} \rangle \lesssim 10\%$ [25].

2.4 BOSON-EXCHANGE POTENTIALS

An accurate account of deuteron properties and the available NN scattering data can also be obtained within the one-boson exchange (OBE) model and its extensions, which can be seen as a straightforward generalization of Yukawa's approach..

In addition to the long-range OPE component, the Bonn potential [26] includes exchange of the isoscalar vector meson ω , with mass $m_\omega = 782$ MeV, and the isovector vector meson ρ, whit mass $m_\rho = 770$ MeV, driving the short range behaviour. The intermediate range attraction originates from exchange of a fictitious scalar isoscalar meson dubbed σ , with mass $m_\sigma = 550$ MeV, which can be interpreted as the exchange of two correlated pions.

The adjustable parameters in the Bonn potential are the meson coupling constants and the cutoffs, $1 \lesssim \Lambda \lesssim 2$ GeV, determining the range of the monopole form factors associated with each meson-nucleon vertex.

An improved version of the Bonn potential, referred to as Charge-Dependent (CD) Bonn potential [27], takes into account the effects of charge-simmetry and charge-independence breaking. In addition, in the scalar-isoscalar channel it includes two mesons, referred to as σ_1 and σ_2, describing correlated two-pion and pion-omega exchange, respectively.

2.5 POTENTIALS BASED ON CHIRAL LAGRANGIANS

In the 1990s, S. Weinberg suggested a way to derive the nuclear potentials from an effective Lagrangian, involving pions and low-momentum nucleons, constrained by the broken chiral symmetry of strong interactions [28]. This approach provides a systematic scheme, referred to as Chiral Effective Field Theory (χEFT), in which the nuclear interaction is expanded in powers of a small parameter, e.g. the ratio between the pion mass or the nucleon momentum, Q, and the scale of chiral symmetry breaking, $\Lambda_\chi \sim 1$ GeV. Within this framework, pion exchange accounts for long- and intermediate-range nuclear forces, whereas short-range interactions are described by contact terms. The main advantage of χEFT lies in the ability to derive two- and many-nucleon potentials from a unified formalism. Moreover, it allows, at least in principle, to improve the potentials with the inclusion of higher order terms, and estimate the theoretical uncertainty.

At leading order (LO), the only contributing mechanism is one-pion exchange, and the NN interaction is fully determined by measured properties of the π-nucleon system. At higher orders—next-to-leading order (NLO), next-to-next-to-leading order (N^2LO), ...—the potential also involves a set of unknown coefficients associated with the contact terms, dubbed low-energy constants, or LECs, to be determined by fitting two-nucleon data. Three-nucleon forces, appearing at N^2LO, depend on two additional LECs, the values of which are usually adjusted to reproduce the binding energies of the three- and four-nucleon systems.

Being based on a momentum expansion, early χEFT potentials were naturally derived in momentum space [29, 30]. However, a procedure has been also developed to obtain coordinate space representations, needed for use in QMC calculations [31, 32]. The numerical results of a study carried out using the Auxiliary Field Diffusion Monte Carlo (AFDMC) technique demonstrate that local coordinate space N^2LO potentials, describing both two- and three-nucleon interactions, provide a remarkably good account of the ground-state energies and charge radii of nuclei with $A \leq 16$ [33].

Because the difference between the mass of the Δ resonance and the nucleon mass, $\Delta m \lesssim 300$ MeV, is small compared to the scale Λ_χ, χEFT can be naturally extended to include explicit Δ degrees of freedom. Fully local, coordinate-space two- and three-nucleon chiral potentials with Δ intermediate states, obtained at N^3LO and N^2LO, respectively, have been shown to explain the energy spectra of nuclei with mass range $A \leq 12$ with a few percent accuracy [34].

Theoretical studies based on χEFT have been also extended to nuclear matter. The present development of the QMC approach only allows to treat pure neutron matter, hereafter referred to as PNM, whereas combined analyses of PNM and SNM have been carried out within the framework of more approximated methods. A prominent feature of the chiral three-nucleon interactions, tuned to reproduce the properties of few-nucleon systems, is the capability to predict saturation of SNM at $\varrho \approx \varrho_0$ [35]. However, the results of AFDMC calculations indicate that in PNM the ambiguities associated with the choice of contact operators are much larger than in light nuclei [36].

It has to be kept in mind that χEFT is based on a low momentum expansion. Therefore, it is inherently limited when it comes to describing nuclear interactions in high-density nuclear matter. This problem clearly emerges from the phase-shift analysis of Piarulli *et al.* [38], showing that, even with the inclusion of significant corrections up to N^3LO, chiral potentials provide an accurate fit to the data only at $E_{\rm lab} \lesssim 200$ MeV. The failure of the potentials of Gezerlis *et al.* [31] to reproduce the $^1S_0$ phase shifts at larger energies is illustrated in Fig. 2.8. Based on the argument discussed in Sect. 2.3.1, Fig. 2.8 suggests that the χEFT potential is unable to accurately describe NN interactions in nuclear matter at $\varrho \gtrsim 1.5\varrho_0$. In the density regime relevant to neutron stars, purely phenomenological interactions, such as the AV18 potential, appear to provide a better option [37].

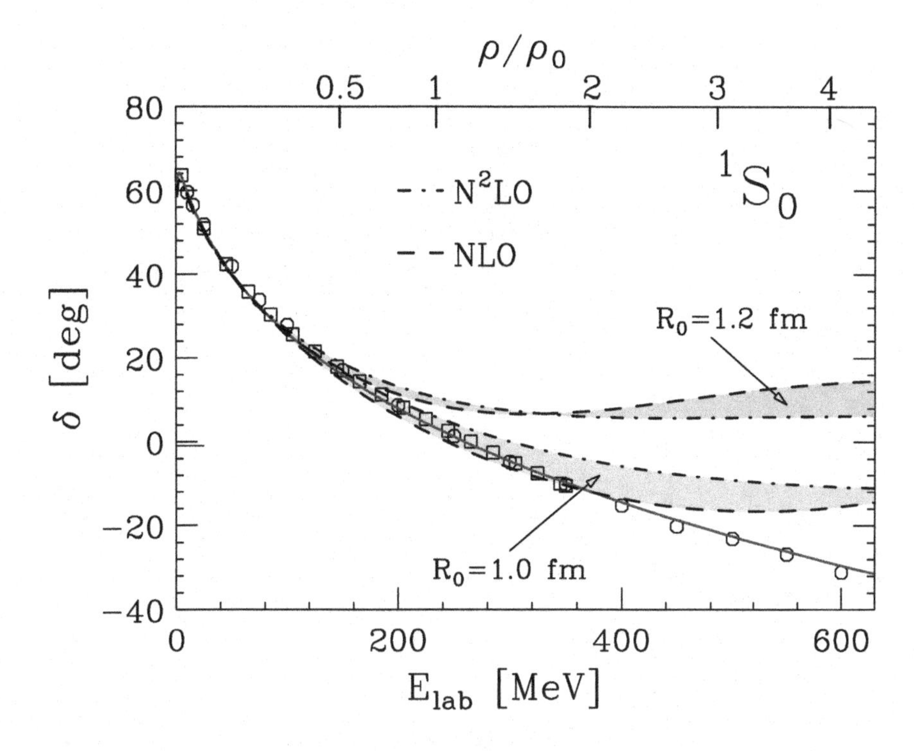

Figure 2.8 Proton-neutron scattering phase shifts in the $^1S_0$ channel, as a function of kinetic energy of the beam particle in the laboratory frame (bottom axis). The corresponding density of PNM, in units of $\varrho_0 = 0.16$ fm^{-3}, is given in the top axis. The shaded regions illustrate the spread between the NLO (dashed lines) and N^{2}LO (dot-dash lines) predictions of the χEFT potentials of Gezerlis *et al.* [31], for two different choices of the cutoff R_0. Squares and circles represent the results of the analyses of the Nijmegen Group [15, 16] and Workman *et al.* [17], respectively. For comparison, the results obtained using the AV18 potential are shown by the solid line. Taken from [37].

CHAPTER 3

NUCLEAR MATTER PROPERTIES

In this chapter, we identify the physical quantities describing the main properties of nuclear matter. The equation of state, encoding the relation between matter pressure and density, plays a critical role in astrophysical applications, in that it determines the mass and radius of stable neutron stars. Single-particle properties, whose unambiguous identification in interacting many-body systems involves non trivial conceptual issues, are also discussed, using the concepts underlying Landau's theory of normal Fermi liquids and the Green's function formalism. Because many essential features of nuclear matter are dictated by translation invariance and Fermi-Dirac statistics only, independent of nuclear dynamics, we also outline the Fermi gas model—in which interactions involving the constituent nucleons are neglected altogether—that will be often used as a pedagogical tool. The calculation of both the equation of state and the Green's function of nuclear matter within the framework of nuclear many-body theory will be discussed in Chapters 4 and 5.

3.1 THE FERMI GAS MODEL

Owing to translation invariance, in nuclear matter one-nucleon states are known to be eigenstates of the momentum operator, labelled by the corresponding eigenvalue $\mathbf{k}$. The associated wave functions can be written in the form

$$\phi_{\mathbf{k}\sigma\tau}(\mathbf{r}) = \varphi_{\mathbf{k}}(\mathbf{r})\chi_\sigma\eta_\tau \ , \tag{3.1}$$

where χ and η are two-component spinors acting in spin and isospin space, respectively, while

$$\varphi_{\mathbf{k}}(\mathbf{r}) = \sqrt{\frac{1}{V}}\ e^{i\mathbf{k}\cdot\mathbf{r}} \ , \tag{3.2}$$

$V = L^3$ being the volume of the normalisation box. For finite V, the discrete spectrum of momentum eigenvalues is determined imposing the periodic boundary conditions

$$\varphi_{\mathbf{k}}(x, y, z) = \varphi_{\mathbf{k}}(x + n_x L, y + n_y L, z + n_z L) \ , \tag{3.3}$$

with $\mathbf{r} \equiv (x, y, z)$ and $n_x, n_y, n_z = 0, \pm 1, \pm 2, \ldots$. The above equation obviously implies the relations

$$k_x = \frac{2\pi n_x}{L}, \quad k_y = \frac{2\pi n_y}{L}, \quad k_z = \frac{2\pi n_z}{L} \ , \tag{3.4}$$

with $\mathbf{k} \equiv (k_x, k_y, k_z)$, yielding the momentum eigenvalues.

In order to simplify the notation, hereafter the spin and isospin indices will be omitted, unless necessary, and the single-nucleon wave function of Eq.(3.1) will be denoted $\phi_{\mathbf{k}}$.

In a non interacting $\mathcal{N}$-particle system in its ground state—corresponding to zero temperature—the constituent particles occupy the momentum eigenstates belonging to eigenvalues $\mathbf{k}$ such that $|\mathbf{k}| = (k_x^2 + k_y^2 + k_z^2)^{1/2} \leq k_F$, with k_F being referred to as Fermi momentum

The Fermi momentum can be related to the particle density, $\varrho = \mathcal{N}/V$, using the equation

$$\nu \sum_{|\mathbf{k}| \leq k_F} = \mathcal{N} \ , \tag{3.5}$$

where the factor ν takes into account the spin-isospin degeneracy of momentum eigenstates[1]. In the thermodynamic limit—corresponding to $\mathcal{N}$, $V \to \infty$ and finite ϱ—Eq.(3.5) becomes

$$\nu \frac{V}{2\pi^2} \int_0^{k_F} |\mathbf{k}|^2 dk = \mathcal{N} \ , \tag{3.6}$$

yielding

$$\varrho = \frac{\mathcal{N}}{V} = \nu \frac{k_F^3}{6\pi^2} \ . \tag{3.7}$$

Note that the above result follow from translation invariance and Fermi-Dirac statistics only, and apply to all system of $\mathcal{N}$ fermions at zero temperature, regardless of the occurrence of interactions.

The ground-state energy per particle is obtained from

$$\frac{E_0}{\mathcal{N}} = \frac{1}{\mathcal{N}} \nu \sum_{|\mathbf{k}| \leq k_F} e_k = \frac{1}{\varrho} \nu \int \frac{d^3k}{(2\pi)^3} \theta(k_F - |\mathbf{k}|) \, e_k \ , \tag{3.8}$$

where

$$e_k = \frac{\mathbf{k}^2}{2m} \ , \tag{3.9}$$

m being the constituent mass, is the energy of a particle carrying momentum $\mathbf{k}$. Substituting Eq.(3.9) in Eq.(3.8) and using Eq.(3.7) one finds

$$\frac{E_0}{\mathcal{N}} = \frac{3}{5} \frac{k_F^2}{2m} = \frac{3}{10m} \left(\frac{6\pi^2}{\nu} \right)^{2/3} \varrho^{2/3} = \frac{3}{5} e_F \ , \tag{3.10}$$

with e_F being the Fermi energy.

A complete set of Fermi gas states can be constructed promoting nucleons from hole states, with momentum $|\mathbf{h}| < k_F$, to particle states, with momentum $|\mathbf{p}| > k_F$. Each state is characterised by the distribution $n(\mathbf{k})$, specifying the occupation of momentum eigenstates. The ground state corresponds to

$$n(\mathbf{k}) = n_0(\mathbf{k}) = \theta(k_F - |\mathbf{k}|) \ . \tag{3.11}$$

Note that momentum conservation prohibits the admixture of the Fermi gas ground state $|0\rangle$, having vanishing total momentum, with one-particle–one-hole states $|\mathbf{p}, \mathbf{h}\rangle$, the total momentum of which is $\mathbf{p} - \mathbf{h} \neq 0$.

[1]The spin-isospin degeneracy is $\nu = 2$ and 4 in PNM and SNM, respectively.

3.1.1 Energy-density and pressure of the degenerate Fermi gas

A non interacting Fermi gas at temperature T is said to be degenerate if thermal energies are negligible compared to the typical kinetic energies, that is, if $T \ll e_F$. The expression of the Fermi energy in terms of ϱ can be used to determine the density regime in which degeneracy sets in, corresponding to

$$\varrho \gg \frac{\nu}{6\pi^2} (2mT)^{3/2} . \tag{3.12}$$

Under the above condition, one can safely set $T = 0$, and the energy density of the system, ϵ, can be obtained from Eq.(3.8). One finds

$$\epsilon = \frac{E_0}{V} = \varrho \frac{E_0}{A} = \frac{\nu}{2\pi^2} \frac{k_F^5}{10m} = \frac{3}{10m} \left(\frac{6\pi^2}{\nu} \right)^{2/3} \varrho^{5/3} . \tag{3.13}$$

The Fermi gas pressure can be derived from the thermodynamic definition, see, e.g., [39]

$$P = -\frac{\partial E_0}{\partial V} = \varrho^2 \frac{\partial}{\partial \varrho} \frac{E_0}{\mathcal{N}} , \tag{3.14}$$

with $E_0/\mathcal{N}$ given by Eq.(3.10), yielding

$$P = \frac{1}{5m} \left(\frac{6\pi^2}{\nu} \right)^{2/3} \varrho^{5/3} . \tag{3.15}$$

3.1.2 Transition to the relativistic regime

So far, we have been assuming that the constituents of the degenerate gas behave as non relativistic particles. However, the properties of the system depend primarily on the distribution of quantum states, $n(\mathbf{k})$, and are largely unaffected by this assumption. Releasing the non relativistic approximation simply amounts to replacing the non relativistic energy, Eq.(3.9), with its relativistic counterpart, acording to

$$e_k = \frac{k^2}{2m} \to \sqrt{k^2 + m^2} . \tag{3.16}$$

Note that the above definition includes the contribution arising from the rest mass of the particle.

Substitution of Eq.(3.16) in Eq.(3.8) allows to determine the ground-state energy per particle of the relativistic Fermi gas, which can in turn be employed to obtain the energy density and the pressure. The density dependence of the resulting expressions turns out to be

$$\epsilon \propto \varrho^{4/3} \quad , \quad P \propto \varrho^{4/3} , \tag{3.17}$$

to be compared to the $\propto \varrho^{5/3}$ dependence characteristic of the non relativistic approximation, see Eqs.(3.13) and (3.15).

The transition to the relativistic regime occurs when the nucleon Fermi energy becomes comparable to the particle mass. It is therefore possible to determine a density ϱ_c such that for $\varrho \gg \varrho_c$ the system is fully relativistic. In nuclear matter, the value of ϱ_c can be estimated from

$$\varrho_c = \nu \frac{\sqrt{2}}{3\pi^2} m^3 , \tag{3.18}$$

with $m \approx 939$ MeV ≈ 4.75 fm^{-1} being the nucleon mass[2]. For $\nu = 2$ the above equation yields the result $\varrho_c \gtrsim 60\varrho_0$, thus confirming that in nuclear matter nucleons can always be described as non relativistic particles.

3.1.3 Extension to non-zero temperature

In the non interacting Fermi gas at $T = 0$, all momentum eigenstates belonging to eigenvalues $\mathbf{k}$ such that $|\mathbf{k}| < k_F$ are occupied with unit probability, while those having $|\mathbf{k}| > k_F$ are vacant. The corresponding distribution, $n_0(k)$, is the Heaviside function $\theta(k_F - |\mathbf{k}|)$ appearing in the definition of the ground-state energy per particle, Eq.(3.8).

At non vanishing temperature, $T \neq 0$, all thermodynamic functions can be derived from the gran canonical partition function

$$Z = \text{Tr}\ \Phi \ , \tag{3.19}$$

where, in the case of a one-component system

$$\Phi = e^{-\beta(H - \mu N)} \ . \tag{3.20}$$

In the above equation, $\beta = T^{-1}$, while N and H denote the particle number and Hamiltonian operators, respectively. The chemical potential μ is defined as

$$\mu = \left(\frac{\partial E}{\partial \mathcal{N}} \right)_V \ , \tag{3.21}$$

E and $\mathcal{N}$ being the energy and the average particle number. Note that in the $T = 0$ limit the above equation yields $\mu = e_F$.

The distribution function $n(\mathbf{k})$—needed to carry out calculations of the thermodynamic functions describing the state of the system—is obtained from the minimisation of the gran canonical potential

$$\Omega = -\frac{1}{\beta} \ln Z \ , \tag{3.22}$$

leading to the equation

$$\frac{\delta \Omega}{\delta n} = 0 \ . \tag{3.23}$$

The result is the Fermi distribution

$$n(\mathbf{k}) = [1 + e^{\beta(e_k - \mu)}]^{-1} \ , \tag{3.24}$$

which reduces to $\theta(\mu - e_k)$ in the zero-temperature limit, corresponding to $\beta \to \infty$.

Figure 3.1 displays the momentum dependence of $n(\mathbf{k})$ for degeneracy $\nu = 4$ and Fermi momentum $k_F = 1.33$ fm^{-1}, specifying SNM at density $\varrho = \varrho_0 = 0.16$ fm^{-3}. The dot-dash, dashed and dotted lines, corresponding to $T = 1$, 2, and 5 MeV, respectively, illustrate the departure from the zero-temperature distribution, represented by the solid line. It clearly appears that at $T = 1$ and 2 MeV the system is still nearly degenerate, while significant thermal effect are visible at the highest temperature. These features can be easily explained

[2]In the natural system of units adopted in his book, 1 MeV $= 1/197.3$ fm^{-1}.

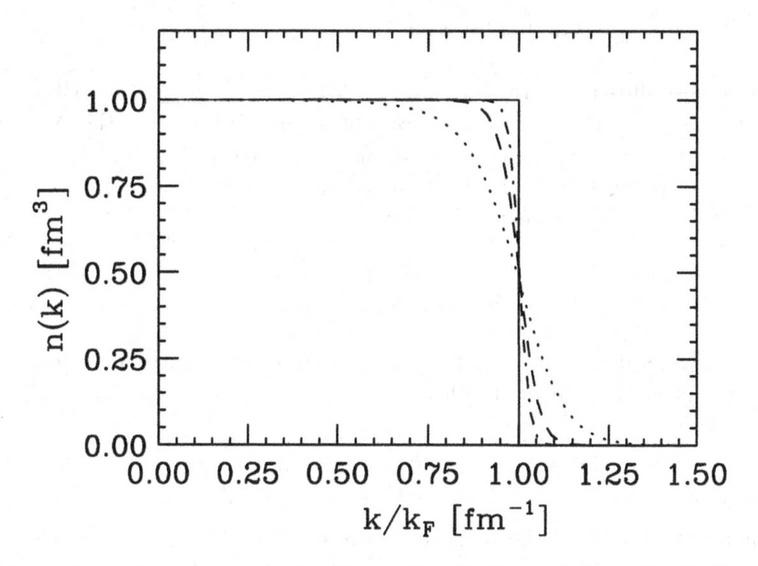

Figure 3.1 Momentum dependence of the Fermi distributions, defined by Eq.(3.24), at density $\varrho = \varrho_0 = 0.16$ fm^{-3} and Fermi momentum $k_F = 1.33$ fm^{-1}. The dot-dash, dashed and dotted lines correspond to $T = 1$, 2, and 5 MeV, respectively. For comparison, the $T = 0$ distribution is also shown by the solid line.

considering that the the condition of near degeneracy is $T/\mu \ll 1$, and the zero-temperature chemical potential at density $\varrho = \varrho_0$ turns out to be $\mu = e_F \approx 22$ MeV.

In neutron stars, the density largely exceeds ϱ_0, and typical temperatures are of the order of 10^6 K $= 8.62 \times 10^{-5}$ MeV. Therefore, in most instances neutron star matter can be treated as a fully degenerate system.

Note that the above discussion applies to both PNM and SNM. In PNM there is only one chemical potential, while in SNM the proton and neutron chemical potentials, μ_p and μ_n, coincide. However, the extension to the case of matter with arbitrary proton fraction, in which $\mu_p \neq \mu_n$, does not involve any conceptual difficulties.

3.2 THE EQUATION OF STATE

The equation of state (EOS) is a nontrivial relation linking the thermodynamic variables that characterise the state of a physical system [39]. The archetypal example is Boyle's ideal gas law, stating that the pressure of a collection of $\mathcal{N}$ noninteracting, pointlike classical particles enclosed in a volume V, grows linearly with the temperature T and the average particle density $\varrho = \mathcal{N}/V$.

The ideal gas law provides a good description of very dilute systems. In general, the EOS can be written as an expansion of the pressure, P, in powers of the density

$$P = \varrho T \left[1 + \varrho B(T) + \varrho^2 C(T) + \ldots\right] . \tag{3.25}$$

The coefficients appearing in the right-hand side of the above equation, which goes under the name of virial expansion, are functions of temperature only. They describe the deviations from the ideal gas law, and, in principle, can be calculated in terms of the underlying elementary interactions. It follows that the EOS carries a great deal of dynamical information, and provides a link between measurable *macroscopic* quantities, such as pressure or energy density, and the forces acting between the constituents of the system at *microscopic* level.

This point is best illustrated by the van der Waals EOS, which describes a collection of classical particles interacting through a potential—reminiscent of the NN potential of Fig. 2.5—featuring a strong repulsive core and an attractive tail, as illustrated in Fig. 3.2. For $|U_0|/T \ll 1$, U_0 being the strength of the attractive component of the potential, the van der Waals EOS reduces to the simple form

$$P = \frac{\varrho T}{1 - \varrho b} - a\varrho^2 \ , \tag{3.26}$$

and the two quantities a and b, taking into account interaction effects, can be directly related to the potential $v(r)$ through [40]

$$a = \pi \int_{2r_0}^{\infty} |v(r)|^2 \ r^2 dr \quad , \quad b = \frac{16}{3}\pi r_0^3 \ , \tag{3.27}$$

where $2r_0$ denotes the radius of the repulsive core, see Fig. 3.2.

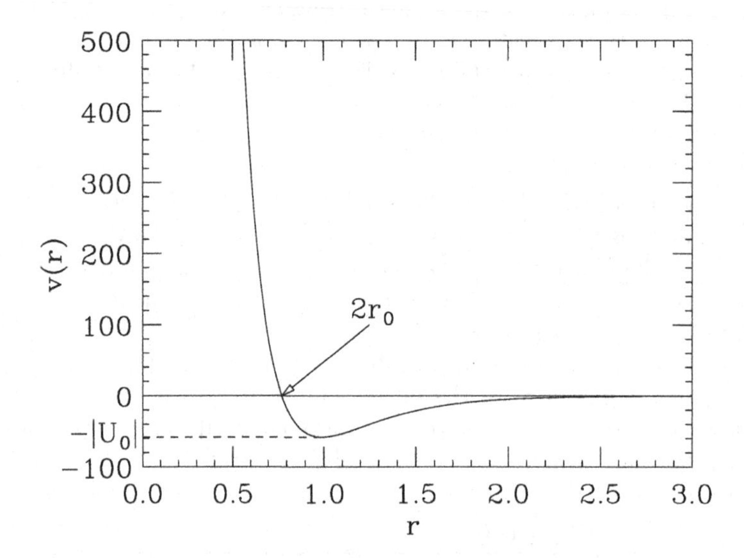

Figure 3.2 Behavior of the potential describing the interactions between constituents of a van der Waals fluid. The particle separation distance r and $v(r)$ are both given in arbitrary units.

3.2.1 Equation of state of cold nuclear matter

The body of data on nuclear masses provides information on the ground-state energy per particle of nuclear matter at $T = 0$, $e(\varrho) = E_0(\varrho)/A$, and can be exploited to constrain the nuclear matter EOS.

The empirical equilibrium properties of isospin-symmetric matter can be inferred from the semi empirical mass formula (1.7), yielding

$$e_0 = e(\varrho = \varrho_0, T = 0) = -16 \text{ MeV}/A \quad , \quad \varrho_0 \approx 0.16 \text{ fm}^{-3} \ . \tag{3.28}$$

In the vicinity of the equilibrium density $e(\varrho)$ can be expanded according to

$$e(\varrho) \approx e_0 + \frac{1}{2}\frac{K}{9}\frac{(\varrho - \varrho_0)^2}{\varrho_0^2} , \tag{3.29}$$

where

$$K = 9\ \varrho_0^2 \left(\frac{\partial^2 e}{\partial \varrho^2}\right)_{\varrho=\varrho_0} = 9 \left(\frac{\partial P}{\partial \varrho}\right)_{\varrho=\varrho_0} , \tag{3.30}$$

is the compressibility module, the value of which can be determined from measurements of the compressional modes of atomic nuclei. The authors of [41, 42] report the result $K = 240 \pm 20$ MeV.

The quadratic extrapolation of Eq.(3.29) cannot be expected to work far from equilibrium density. Assuming a parabolic behavior of $e(\varrho)$ at $\varrho \gg \varrho_0$ leads in fact to predict a speed of sound in matter, v_s, larger than the speed of light, that is

$$v_s = \left(\frac{\partial P}{\partial \epsilon}\right) = \frac{1}{\varrho}\left(\frac{\partial P}{\partial e}\right) > 1 , \tag{3.31}$$

regardless of the value of K. Equation (3.31), where $\epsilon = E_0/V$ denotes the energy density, shows that causality requires

$$\left(\frac{\partial P}{\partial \epsilon}\right) < 1 . \tag{3.32}$$

For a noninteracting Fermi gas in the relativistic regime, see Eq.(3.17), $\epsilon \propto \varrho^{4/3}$, implying

$$P \leq \frac{\epsilon}{3} \quad , \quad v_s \leq \frac{1}{3} , \tag{3.33}$$

where the equal sign corresponds to the limit of massless particles. In the presence of interactions, however, the above limits can be easily exceeded. For example, modelling the repulsion between nucleons in terms of a hard core obviously leads to predict infinite pressure at finite density.

The stiffest EOS compatible with causality is $P = \epsilon$, yielding $v_s = 1$. In the early 1960s, Ya. Zel'dovich demonstrated that the $v_s = 1$ limit, corresponding to $\epsilon \propto \varrho^2$, is indeed attained in a simple semi realistic model of nuclear dynamics, in which nucleons are assumed to interact through exchange of a vector meson [43].

3.2.2 Symmetry energy

The ground-state energy per nucleon of matter with proton and neutron densities $\varrho_p = x\,\varrho$ and $\varrho_n = (1-x)\varrho$, respectively, can be expanded in series of powers of the quantity $\delta = 1 - 2x = (\varrho_n - \varrho_p)/\varrho$, providing a measure of neutron excess. The resulting expression reads

$$\frac{1}{A}E_0(\varrho,\delta) = \frac{1}{A}E_0(\varrho,0) + E_{\rm sym}(\varrho)\delta^2 + O(\delta^4) , \tag{3.34}$$

where the symmetry energy

$$\begin{aligned} E_{\rm sym}(\varrho) &= \left\{ \frac{\partial[E(\varrho,\delta)/A]}{\partial\delta^2} \right\}_{\delta=0} \\ &\approx \frac{1}{A}E(\varrho,1) - \frac{1}{A}E_0(\varrho,0) , \end{aligned} \tag{3.35}$$

can be interpreted as the energy required to convert SNM into PNM. The density dependence of $E_{\rm sym}(\varrho)$, that can be obtained expanding around the equilibrium density of SNM, ϱ_0, is conveniently characterised by the quantity

$$L = 3\varrho_0 \left(\frac{dE_{\rm sym}}{d\varrho} \right)_{\varrho=\varrho_0} . \tag{3.36}$$

Empirical information on both $E_{\rm sym}(\varrho_0)$ and L have been extracted from data collected by both laboratory experiments and astrophysical observations. The survey of 28 analyses, carried out by B.-A. Li and X. Han, reports the results $E_{\rm sym}(n_0) = 31.6 \pm 2.66$ and $L = 58.9 \pm 16$ MeV [44] .

3.2.3 Pressure

Knowing the density dependence of the ground-state energy, $E_0(\varrho)$, the pressure of nuclear matter—which plays a critical role in the determination of the maximum mass of neutron stars—can be obtained from the thermodynamic definition, Eq.(3.14).

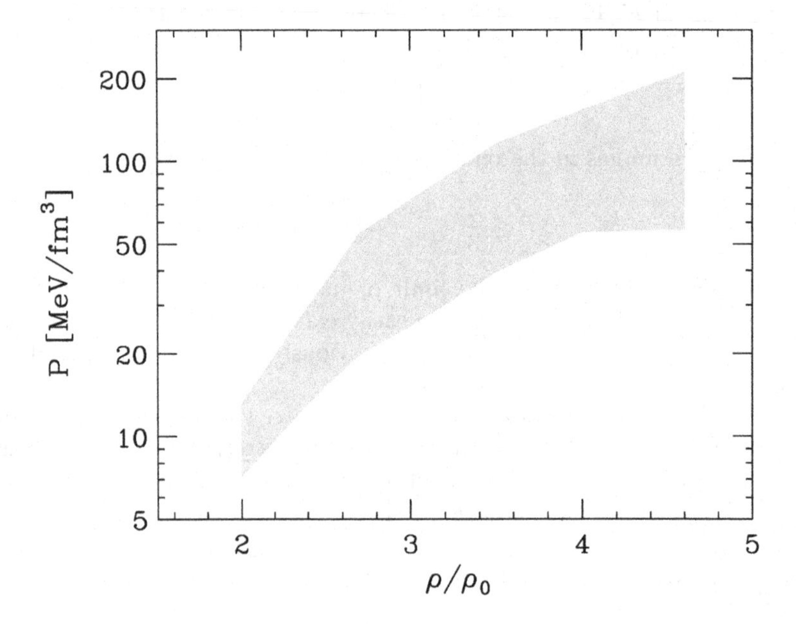

Figure 3.3 Density dependence of the pressure of SNM. The shaded area corresponds to the region compatible with the results of the analysis of Danielewicz *et al.* [45].

At equilibrium density, the pressure, P, is trivially related to the compressibility module, see Eq.(3.30). Empirical information on its density dependence have been obtained through the analysis of data from nuclear collision. These processes lead to the formation of compressed nuclear matter—with density largely exceeding nuclear saturation density—which then expands to return to equilibrium conditions. Therefore, information on $P(\varrho)$ at $\varrho > \varrho_0$ can be extracted from the study of matter flow.

The region of the pressure-density plane compatible with the results of the analysis of the flow data carried out by Danielewicz *et al.* [45] is shown in Fig. 3.3. Note that the density is given in units of $\varrho_0 = 0.16$ fm^{-3}.

3.3 SINGLE–NUCLEON PROPERTIES

The conceptual framework for the identification of single–particle properties in interacting many-body systems is laid down in Landau's theory of normal Fermi liquids [46]. Here the basic tenet is that, if interactions can be switched on adiabatically, there is a one-to-one correspondence between the eigenstates of the interacting system and the Fermi gas eigenstates, which are described by the same distribution function. Note that this fundamental property is what defines a normal Fermi liquid, and is not fulfilled by, for example, superconducting or superfluid systems.

If a particle of momentum $\mathbf{k}$ such that $|\mathbf{k}| > k_F$ is added to the Fermi gas ground state and the interaction is turned on, the resulting state, characterised by the distribution $n = n_0 + \delta n$, comprises the ground state of $\mathcal{N}$ interacting particles plus a particle whose properties—most notably the mass—are modified by the presence of the surrounding medium. This dressed particle is referred to as a quasiparticle.[3].

The energy of a state is a functional of the distribution function, $E[n]$, the expression of which is, in general, very cumbersome. However, it can be conveniently expanded in series of powers of the difference $\delta n = n - n_0$. The resulting expression

$$E[n] = E[n_0] + \sum_k e_k \delta n(\mathbf{k}) + O[\delta n^2(\mathbf{k})] , \tag{3.37}$$

shows that if the distribution n is close to the ground-state distribution, so that terms of order δn^2 can be neglected, the first functional derivative of E with respect to n, evaluated for $n = n_0$, can be identified with the energy of a quasiparticle carrying momentum $\mathbf{k}$.

Within the above picture, the energy of a quasiparticle on the Fermi surface can be obtained by adding a particle of momentum $k = k_F$ to the $\mathcal{N}$-particle system, leaving the volume V unchanged. The $\mathcal{N}$- and $(\mathcal{N}+1)$-particle systems are both in the ground state, and, in the $\mathcal{N} \to \infty$ limit, one obtains

$$e_{k_F} = e_F = E_0(\mathcal{N}+1) - E_0(\mathcal{N}) \to \left(\frac{\partial E_0}{\partial \mathcal{N}}\right)_V = \left(\frac{\partial}{\partial \varrho} \varrho \frac{E_0}{\mathcal{N}}\right)_V . \tag{3.38}$$

This equation, establishing a relation between the Fermi energy and the ground-state energy, E_0, is one of the few exact results of the theory of interacting many-body systems, known as Hugenholtz-Van Hove (HVH) theorem [47]. In addition, from the thermodynamic definition of the chemical potential, Eq.(3.21), it follows that $\mu = e_F$.

Equation (3.37) can be employed to determine e_k for any values of k. The energies corresponding to $k > k_F$ and $k < k_F$ are obtained by adding or removing a particle of momentum $\mathbf{k}$, respectively, and subtracting the ground-state energy from the energy of the resulting state.

The above discussion shows that the one-to-one correspondence between Fermi gas states and states of the interacting systems is established through the replacement of bare particles with quasiparticles. However, It has to be pointed out that, strictly speaking, this procedure can only be applied to the ground state, which is by definition stable. Excited states of the interacting system, on the other hand, are not stable. Owing to the occurrence of interactions between quasiparticles, described by the quadratic term in Eq.(3.37), they decay with a lifetime $\tau_k \propto 1/|e_k - e_F|^2$. Hence, the concept of quasiparticle is fully justified only in the vicinity of the Fermi surface.

[3]In the following spin and isospin quantum numbers will be omitted, and quasiparticle states will be labelled by the momentum eigenvalue $\mathbf{k}$ only

3.3.1 Green's function and spectral function

The Green's function formalism provides a mathematical framework ideally suited for the description of single-particle properties of normal Fermi liquids. The two-point Green's function, also referred to as propagator, is defined by the equation

$$G(\mathbf{k},\omega) = -i \int dt \ \langle 0|T\{a_{\mathbf{k}}(t)a^{\dagger}_{\mathbf{k}}(0)\}|0\rangle \ e^{i\omega t} \ , \tag{3.39}$$

where $|0\rangle$ denotes the system ground state, $a^{\dagger}_{\mathbf{k}}$ and $a_{\mathbf{k}}$ are creation and annihilation operators of a particle of momentum $\mathbf{k}$, respectively, and

$$T\{a_{\mathbf{k}}(t)a^{\dagger}_{\mathbf{k}}(0)\} = \begin{cases} a_{\mathbf{k}}(t)a^{\dagger}_{\mathbf{k}}(0) & , \ t>0 \\ -a^{\dagger}_{\mathbf{k}}(0)a_{\mathbf{k}}(t) & , \ t<0 \end{cases} , \tag{3.40}$$

with

$$a_{\mathbf{k}}(t) = e^{iHt}a_{\mathbf{k}}e^{-iHt} \ , \tag{3.41}$$

H being the Hamiltonian operator[4].

Equation (3.39) can be rewritten in the form

$$G(\mathbf{k},\omega) = \int dt \ G(\mathbf{k},t) \ e^{i\omega t} \ , \tag{3.42}$$

with

$$G(\mathbf{k},t) = \begin{cases} -ie^{iE_0 t}\langle 0|a_{\mathbf{k}}e^{-iHt}a^{\dagger}_{\mathbf{k}}|0\rangle & , \ t>0 \\ ie^{-iE_0 t}\langle 0|a^{\dagger}_{\mathbf{k}}e^{iHt}a_{\mathbf{k}}|0\rangle & , \ t<0 \end{cases} , \tag{3.43}$$

showing that for positive and negative time, t, the Green's function involves the time evolution of the states $a^{\dagger}_{\mathbf{k}}|0\rangle$ and $a_{\mathbf{k}}|0\rangle$, respectively.

By inserting complete sets of intermediate states in Eq.(3.39) we obtain the expression

$$iG(\mathbf{k},t) = \sum_n \left\{ |\langle n|a^{\dagger}_{\mathbf{k}}|0\rangle|^2 e^{-i(E_n-E_0)t}\theta(t) - |\langle n|a_{\mathbf{k}}|0\rangle|^2 e^{-i(E_0-E_n)t}\theta(-t) \right\} , \tag{3.44}$$

that can be rewritten

$$G(\mathbf{k},t) = \int \frac{d\omega}{2\pi} \ G(\mathbf{k},\omega) \ e^{-i\omega t} \ , \tag{3.45}$$

with

$$G(\mathbf{k},\omega) = \sum_n \frac{|\langle n|a^{\dagger}_{\mathbf{k}}|0\rangle|^2}{\omega - (E_n - E_0) + i\eta} + \sum_n \frac{|\langle n|a_{\mathbf{k}}|0\rangle|^2}{\omega - (E_0 - E_n) - i\eta} , \tag{3.46}$$

where $\eta = 0^+$.

Note that the first and second contribution to the right-hand side of the above equation involve $(\mathcal{N}+1)$- and $(\mathcal{N}-1)$-particle states, respectively. In the absence of interactions, these states are simply obtained from the ground state by adding or removing a particle with momentum $\mathbf{k}$. It follows that

$$\langle n|a^{\dagger}_{\mathbf{k}}|0\rangle = \theta(|\mathbf{k}| - k_F) \ , \ \langle n|a_{\mathbf{k}}|0\rangle = \theta(k_F - |\mathbf{k}|) \ , \tag{3.47}$$

[4]The minus sign appearing in the right-hand side of Eq. (3.40) is a consequence of Fermi-Dirac statistics, requiring that the operators $a^{\dagger}_{\mathbf{k}}$ and $a_{\mathbf{k}}$ obey anticommutation rules.

and Eq.(3.46) reduces to form

$$G_0(\mathbf{k},\omega) = \begin{cases} \dfrac{1}{\omega - \omega_k + i\eta} & , \ |\mathbf{k}| > k_F \\ \\ \dfrac{1}{\omega - \omega_k - i\eta} & , \ |\mathbf{k}| < k_F \end{cases} , \tag{3.48}$$

showing that the Green's function exhibits a simple pole at $\omega = \mathbf{k}^2/2m$.

Within the framework of Landau's theory, the above discussion can also be applied to the case of interacting Fermi systems, replacing the kinetic energy spectrum with the spectrum of quasiparticle energies. However, it has to be kept in mind that quasiparticles decay with a characteristic lifetime $\tau_k = \Gamma_k^{-1}$, which becomes vanishingly small only for states in the vicinity of the Fermi surface.

The general expression of the Green's function of a normal Fermi liquid is

$$G(\mathbf{k},\omega) = \frac{1}{\omega - \omega_k - \Sigma(\mathbf{k}, E)} , \tag{3.49}$$

where $\Sigma(\mathbf{k}, E)$ is a well-behaved complex function—referred to as irreducible, or proper, self-energy—embodying all interaction effects. Using the relation $G_0^{-1}(\mathbf{k},\omega) = \omega - \omega_k$, the above equation can be rewritten as an integral equation for G, of the form

$$G(\mathbf{k},\omega) = G_0(\mathbf{k},\omega) + G_0(\mathbf{k},\omega)\Sigma(\mathbf{k},E)G(\mathbf{k},\omega) , \tag{3.50}$$

referred to as Dyson's equation[5].

In the limit of small Γ_k, in which the quasiparticle concept is fully meaningful, the pole of the Green's function, corresponding to the quasiparticle energy e_k, is obtained by solving the equation

$$e_k = \omega_k + \mathrm{Re}\Sigma(k, e_k) , \tag{3.51}$$

and Eq.(3.49) reduces to

$$G(\mathbf{k},\omega) = \frac{Z_k}{\omega - e_k + i\Gamma_k} , \tag{3.52}$$

with

$$\Gamma_k = Z_k \mathrm{Im}\Sigma(k, e_k) \quad , \quad Z_k = \left[1 + \frac{\partial}{\partial\omega}\mathrm{Re}\Sigma(k,\omega)\right]_{\omega=e_k} . \tag{3.53}$$

The spectral function, $P(\mathbf{k},\omega)$, is trivially related to the Green's function through

$$P(\mathbf{k},\omega) = \mp\frac{1}{\pi}\mathrm{Im}G(\mathbf{k},\omega) . \tag{3.54}$$

Choosing the upper or lower sign, the right-hand side of the above equation describes the probability that by adding to or removing from the $\mathcal{N}$-particle ground state a particle of momentum $\mathbf{k}$, the resulting $(\mathcal{N}+1)$- or $(\mathcal{N}-1)$-particle system is left with energy $E_0 + \omega$ or $E_0 - \omega$, respectively.

---

[5]Note that Eqs.(3.49) and (3.50) are equivalent only in the case in which $G_0(\mathbf{k},\omega)$ and $\Sigma(\mathbf{k},\omega)$ are diagonal in the indices associated with the discrete quantum numbers.

Finally, it has to be pointed out that the Green's function can also be employed to obtain the ground-state energy from

$$E_0 = -\frac{i}{4\pi}\,\nu \sum_{\mathbf{k}} \int_C d\omega\; (\omega + e_k)\; G(\mathbf{k}, \omega)\;, \tag{3.55}$$

where the contour C comprises the real axis and an infinite semicircle in the upper half-plane [48].

CHAPTER 4

NUCLEAR MATTER THEORY

The ultimate goal of the many-body theory of nuclear matter, clearly stated in the 1970s in a seminal paper by H. Bethe [49], is the derivation of its properties from a microscopic description of the underlying dynamics.

Within the scheme outlined in Chapters 2 and 3, the first step towards the achievement of this goal is the determination of the EOS, which amounts to solving the Schrödinger equation to obtain the ground-state wave function and energy of nuclear matter at given density $\varrho = A/V$,

$$H|\Psi_0\rangle = E_0|\Psi_0\rangle \ , \tag{4.1}$$

with H being the Hamiltonian of Eq. (2.1). Owing to the complexity of the nuclear potentials discussed in Chapter 2, however, even this limited problem involves severe difficulties, and its treatment has required the development of highly sophisticated theoretical methods, as well as of suitable approximation schemes.

In this chapter, we highlight the strength and limitations of the dynamical model built on the concept of mean field, and discuss the approaches based on many-body theory that have been most extensively used to study the properties of SNM and PNM at zero temperature.

Following the historical development of nuclear matter theory, we will first focus on G-matrix perturbation theory and on the alternative variational approach based on the formalism of correlated wave functions, which will be further discussed in Chapter 5. The Coupled Cluster (CC) and Self-Consistent Green's Function (SCGF) methods, as well as the approach based on QMC techniques, will also be outlined.

Although the discussion of Section 3.1.2 strongly suggests that nuclear matter predominantly behaves as a non relativistic system, the occurrence of relativistic effects and the impact of the associated corrections on the properties of high-density matter have been studied within different approaches; that will be briefly described in the concluding section.

4.1 THE MEAN-FIELD APPROXIMATION

The mean-field approximation (MFA) is based on the assumption that nucleons can be treated as independent particles, moving in an average field described by the potential U_i.

Formally, this assumption amounts to replacing

$$\sum_{j>i} v_{ij} + \sum_{k>j>} V_{ijk} \to \sum_i U_i \; , \tag{4.2}$$

in the right-hand side of Eq.(2.1). As a result, the nuclear Hamiltonian reduces to the simple form

$$H = \sum_i \mathcal{H}_i \; , \tag{4.3}$$

with

$$\mathcal{H}_i = \frac{\mathbf{p}_i^2}{2m} + U_i \; , \tag{4.4}$$

and the normalised A-nucleon ground state can be written in terms of eigenstates of the single-particle hamiltonian, $\mathcal{H}_i$, according to

$$|\Psi_0\rangle = \frac{1}{\sqrt{A!}} \mathcal{A} \prod_{\alpha_i \in \{F\}} \phi_{\alpha_i}(i). \tag{4.5}$$

Here the product is extended to the A lowest energy eigenstates of $\mathcal{H}_i$—specified by a set of quantum numbers collectively denoted by α_i—comprising the Fermi sea $\{F\}$. Antisymmetry of the nuclear wave function under particle exchange, dictated by Pauli's exclusion principle, is realised through the action of the operator $\mathcal{A}$.

From the above equations, it follows that within the MFA the many-body Schrödinger equation (4.1) reduces to the set of one-body equations

$$\mathcal{H}_i \phi_{\alpha_i}(i) = \epsilon_{\alpha_i} \phi_{\alpha_i}(i) \; , \tag{4.6}$$

with $i = 1, \ldots,$ A, and the ground-state energy is given by

$$E_0 = \sum_{\alpha_i \in \{F\}} \epsilon_{\alpha_i} \; . \tag{4.7}$$

In nuclear matter, translation invariance requires that the potential associated with the mean field be diagonal in momentum space. As a consequence the single-nucleon states are still given by the Fermi gas eigenstates of Eqs.(3.1)-(3.2), and all properties of the system can be obtained by modifying the energy spectrum according to

$$e_{\mathbf{k}} = \frac{\mathbf{k}^2}{2m} \to \frac{\mathbf{k}^2}{2m} + U_{\mathbf{k}} \; . \tag{4.8}$$

Note that, in the language of Landau's theory of normal fermi liquids discussed in Section 3.3, the above replacement amounts to replacing the bare particles with quasiparticles distributed according to the Fermi gas distribution of Eq.(3.11)—with the Fermi momentum given by $k_F = (6\pi^2 \varrho/\nu)^{1/3}$—in the $Z_k \to 1$, $\Gamma_k \to 0$ limit, see Eq.(3.53). Within this picture, interaction effects can be conveniently described introducing an effective mass $m^\star$, playing the role of quasiparticle mass, defined through

$$\frac{1}{m^\star} = \frac{1}{k} \frac{de_{\mathbf{k}}}{dk} \; , \tag{4.9}$$

with $e_{\mathbf{k}}$ given by Eq.(4.8).

4.1.1 Limits of the mean-field approximation

The MFA provides the basis of the nuclear shell model, yielding a remarkably accurate description of a large body of measured nuclear properties, such as the energy spectra and the magnetic moments. However, significant deviations from the shell model predictions have been observed in the cross section of the electron-nucleus scattering processes

$$e + \mathrm{A} \to e' + p + (\mathrm{A} - 1) \ , \tag{4.10}$$

in which the knocked out nucleon and the outgoing electron are detected in coincidence, and the recoiling (A − 1)-nucleon system is left in any bound or continuum states. Accurate measurements of the $(e, e'p)$ cross section have unambiguously demonstrated that, while the spectroscopic lines corresponding to knock out of nucleons occupying the shell-model orbitals are clearly visible in the missing energy spectra, the associated spectroscopic factors—that is, the normalisations of the shell model states—are considerably less than unity, regardless of the nuclear mass number A. Complementary $(e, e'p)$ data, collected in a different kinematical regime, strongly suggest that this feature is a clear-cut manifestation of dynamical NN correlations [50].

Correlations arise from virtual collisions among nucleons in the nuclear medium, leading to the excitation of the participating particles to states with energy larger than the Fermi energy, and to a corresponding depletion of the shell model states belonging to the Fermi sea.

The accuracy of the MFA and the role of correlation effects in nuclei are long standing issues of Nuclear Physics. In their classic book "Theoretical nuclear physics", first published in the 1950s, J. Blatt and V. Weisskopf warned the reader that "The limitation of any independent particle model lies in its inability to encompass the correlation between the positions and spins of the various particles in the system" [51].

Correlation effects can be best identified in translationally invariant systems, and have been also extensively studied in fermionic quantum fluids other than nuclear matter, such as liquid helium [52]. The fundamental element of this analysis is the momentum distribution, which can be written in terms of the Green's function $G(\mathbf{k}.\omega)$ of eq.(3.49) according to [48]

$$n(\mathbf{k}) = -i \int_{\mathcal{C}} \frac{d\omega}{2\pi} \, G(\mathbf{k}.\omega) \ , \tag{4.11}$$

where the integration contour $\mathcal{C}$ consists of the real axis and a semicircle of infinite radius in the upper half plane. The momentum distribution, describing the probability that a particle carry momentum $k = |\mathbf{k}|$, exhibits a discontinuity at $k = k_F$, as shown in Fig. 4.1. In the non interacting system, as well as in the case of interactions that can be described within the MFA, $n(\mathbf{k})$, normalised in such a way as to satisfy

$$\frac{\nu}{\varrho} \int \frac{d^3k}{(2\pi)^3} \, n(\mathbf{k}) = 1 \ , \tag{4.12}$$

reduces to the Fermi gas ground-state distribution

$$n(k) = \theta(k_F - k) \ , \tag{4.13}$$

In the presence of correlations, on the other hand, the momentum distribution can be conveniently written in the form [53]

$$n(k) = Z_k \theta(k_F - k) + n_c(k) \ , \tag{4.14}$$

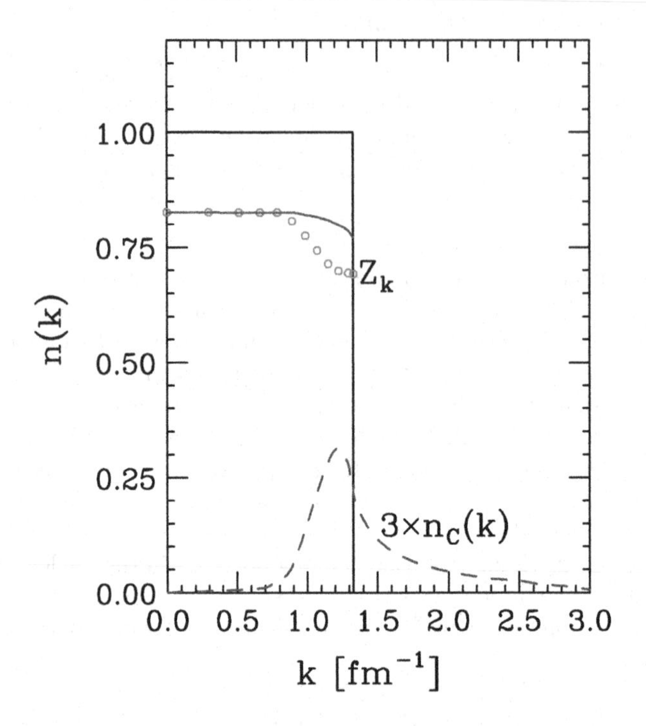

Figure 4.1 The open circles and the dashed line represent the discontinuous (Z_k) and continuous [$3\times n_c(k)$] contributions to the nucleon momentum distribution of SNM at equilibrium density, $n(k)$, displayed by the solid line. The results have been obtained within the framework of nuclear many-body theory, using a phenomenological Hamiltonian including two- and three-nucleon potentials. For comparison, the Fermi gas momentum distribution, $n(\mathbf{k}) = \theta(k_F - k)$, is also shown. Adapted from [53].

where the spectroscopic factor $Z_k < 1$ is defined in Eq.(3.53), and the smooth function $n_c(p)$, describing correlation effects, is continuous across the Fermi surface.

The momentum dependence of $n(k)$, Z_k, and $n_c(k)$ in SNM at equilibrium density are illustrated in Fig. 4.1. It is apparent that, while the most prominent correlation effect is the presence of a high-momentum tail, the contribution of correlations at $k < k_F$ is also significant. In fact, it turns out to account for $\sim$ 25% of the normalisation of $n_c(k)$. Obviously, a non vanishing $n_c(k)$ can only be obtained from a dynamical model including interaction effects beyond the MFA.

4.2 RENORMALISATION OF THE NN INTERACTION

A prominent difficulty involved in performing calculations of nuclear matter properties based on the full nuclear Hamiltonian of Eq. (2.1) lies in the presence of the strongly repulsive core in the NN potential. Owing to this feature, clearly illustrated in Fig. 2.5, the matrix elements of the potential in the basis of eigenstates of the Hamiltonian of the non interacting system—that is, Slater determinants built from the Fermi gas states defined by Eqs.(3.1) and (3.2)—are large, and standard perturbation theory is not applicable.

Two main routes have been historically followed to overcome the above problem. The first one, originally proposed by K. Brueckner in the 1950s, is based on the replacement of the bare nucleon-nucleon potential with a well-behaved effective interaction, describing NN scattering in the nuclear medium, suitable for use in perturbative calculations [49, 54]. In the alternative approach, the Fermi gas basis is replaced by a different set of states, determined from a variational calculation of the ground-state energy. The new basis states, embodying the effects of NN correlations arising from non perturbative interactions, allow to obtain a variety of nuclear matter properties using the bare nuclear Hamiltonian and the formalism of non-orthogonal perturbation theory [55, 56].

More recently, it has been suggested that effective interactions suitable for perturbative calculations can also be obtained combining potentials derived within χEFT and renormalisation group evolution to low momentum. However, in view of the discussion of Chapter 2, the applications of this approach appear to be confined to a rather narrow density region [57, 58, 59].

A conceptually similar treatment, based on the formalism of correlated basis functions and applicable to *any* NN potentials, has been also developed [60]. Within this approach, renormalisation—resulting in the screening of the repulsive core—is carried out in coordinate space, and the quantity driving the evolution is nuclear matter density.

Finally, it has to be emphasised that over the past two decades the development of Quantum Monte Carlo techniques has made it possible to carry out highly accurate calculations of a variety of properties of nuclei with mass number up to A=12 using state-of-the-art nuclear Hamiltonians, see Fig. 2.7. This approach, appears to be very promising, although its applications to infinite system are still limited to PNM [22], and may also provide valuable information, useful to assess the accuracy of more approximate methods.

4.3 *G*-MATRIX PERTURBATION THEORY

The theoretical framework known as Brueckner-Bethe-Goldstone (BBG) theory is based on a perturbative expansion in which the nuclear Hamiltonian is split in the form

$$H = H_0 + H_1 \ , \tag{4.15}$$

with

$$H_0 = \sum_{i=1}^{A} \left[\frac{\mathbf{k}_i^2}{2m} + U_i \right] \ , \tag{4.16}$$

$$H_1 = \sum_{j>i=1}^{A} v_{ij} - \sum_{i=1}^{A} U_i \ , \tag{4.17}$$

where the single particle potential U_i—which in principle does not affect the results of calculations of any physical quantities—is chosen in such a way as to improve the convergence of perturbative calculations [54], or to fulfil specific analytical properties [61]. Note that the Hamiltonian defined by the above equations does not include the three-nucleon potential. The treatment of three-body forces in BBG theory will be discussed at a later stage.

Using the basis of eigenstates of H_0 and the corresponding eigenvalues, provided by the solution of the Schrödinger equations

$$H_0 |\Phi_n\rangle = \mathcal{E}_n |\Phi_n\rangle, \tag{4.18}$$

the ground-state energy of the interacting system, E_0 can be computed using the perturbative expansion in powers of H_1 derived by J. Goldstone [62] in the 1950s. The resulting expression can be cast in the form

$$E_0 = \mathcal{E}_0 + \Delta E \;, \tag{4.19}$$

with

$$\begin{aligned}\Delta E &= \langle \Phi_0 | \sum_{N=0}^{\infty} H_1 \left(\frac{Q}{\mathcal{E}_0 - H_0} H_1 \right)^N | \Phi_0 \rangle_L \\ &= \langle \Phi_0 | H_1 | \Phi_0 \rangle + \langle \Phi_0 | H_1 \frac{Q}{\mathcal{E}_0 - H_0} H_1 | \Phi_0 \rangle \\ &\quad + \langle \Phi_0 | H_1 \frac{Q}{\mathcal{E}_0 - H_0} H_1 \frac{Q}{\mathcal{E}_0 - H_0} H_1 | \Phi_0 \rangle + \dots \;,\end{aligned} \tag{4.20}$$

where the projection operator

$$Q = \sum_{n \neq 0} | \Phi_n \rangle \langle \Phi_n | \;, \tag{4.21}$$

accounts for the effect of Pauli's principle, preventing the state $|\Phi_0\rangle$ from appearing as an intermediate state in Eq.(4.20).

The energy shift $\Delta E = E_0 - \mathcal{E}_0$, can be conveniently represented by diagrams, as illustrated in Fig. 4.2, and the notation $\langle \Phi_0 | \dots | \Phi_0 \rangle_L$ indicates that only linked diagrams must be considered. A dashed line joining two vertices depicts the NN potential, v_{ij}, while a cross corresponds to the insertion of the one-body potential, U_i, and oriented lines labelled with Latin and Greek letters represent the Green's functions describing nucleons in hole and particle states, respectively.

The explicit expression of the first-order contribution in v is given by

$$\Delta E = \frac{1}{2} \sum_{\mathbf{h}_1, \mathbf{h}_2 \in \{F\}} \langle \mathbf{h}_1 \mathbf{h}_2 | v | \mathbf{h}_1 \mathbf{h}_2 \rangle - \langle \mathbf{h}_1 \mathbf{h}_2 | v | \mathbf{h}_2 \mathbf{h}_1 \rangle \;, \tag{4.22}$$

where the sum is extended to all two-nucleon states with momenta $\mathbf{h}_1$ and $\mathbf{h}_2$ belonging to the Fermi sea. The diagrams representing the right-hand side of Eq. (4.22) involve two independent hole lines, and the corresponding contribution to ΔE, which turns out to be proportional to $A\varrho$, is referred to as a two-hole-line contribution. Terms of order $A\varrho$ appear at every order of the perturbative expansion, with the N-th order term comprising contributions proportional to $A\varrho$, $A\varrho^2$, ..., $A\varrho^{N-1}$.

From the formal point of view, Eqs.(4.19) and (4.20) provide a well defined scheme to obtain the ground state energy of nuclear matter from a given microscopic model of the nuclear Hamiltonian. As pointed out in Section 4.2, however, the matrix elements involving the NN potential cannot be treated as small quantities, and low-order approximation to E_0 cannot be expected to work.

This problem may be circumvented by rearranging the perturbative series in such a way as to collect the terms proportional to $A\varrho$ to all orders, and neglecting those involving higher powers of ϱ altogether. As a result, the bare NN potential is replaced by a well-behaved operator, referred to as G-matrix, describing a NN scattering process at energy W taking place in the nuclear medium. The G-matrix is defined by the integral equation

$$G(W) = v + v \frac{Q}{W - H_0} G(W) \;, \tag{4.23}$$

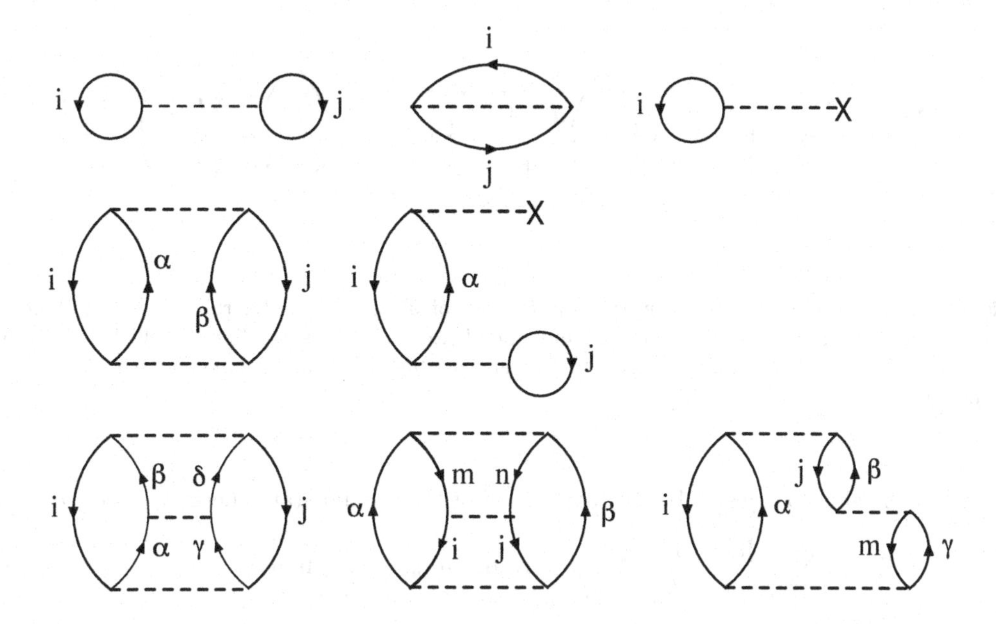

Figure 4.2 Diagrams contributing to the energy shift $\Delta E = E_0 - \mathcal{E}_0$, see Eq.(4.20). Dashed lines joining two vertices and crosses depict the NN potential, v_{ij} and the insertion of the one-body potential, U_i, respectively. Oriented lines labelled with Latin and Greek letters represent the Green's functions describing nucleons in hole and particle states.

with Q given by Eq.(4.21).

The approximation of neglecting all terms proportional to $A\varrho^N$ with $N \geq 2$ is clearly designed to work in the low-density limit. Its applicability to nuclear matter is supported by the argument that, because the range of the repulsive core of the NN potential is short with respect to the average distance between nucleons, in NN collisions nuclear matter behaves largely as a dilute system.

As an example, the expression of the direct matrix element of $G(W)$ between two-nucleon states $|\mathbf{h}_1\mathbf{h}_2\rangle$ with $|\mathbf{h}_1|, |\mathbf{h}_2| < k_F$ is given by

$$\begin{aligned}\langle\mathbf{h}_1\mathbf{h}_2|G(W)|\mathbf{h}_1\mathbf{h}_2\rangle &= \langle\mathbf{h}_1\mathbf{h}_2|v|\mathbf{h}_1\mathbf{h}_2\rangle \\ &+ \sum_{\mathbf{p}_1,\mathbf{p}_2 \notin \{F\}} \frac{\langle\mathbf{h}_1\mathbf{h}_2|v|\mathbf{p}_1\mathbf{p}_2\rangle\langle\mathbf{p}_1\mathbf{p}_2|G(W)|\mathbf{h}_1\mathbf{h}_2\rangle}{W - e_{\mathbf{p}_1} - e_{\mathbf{p}_2}} ,\end{aligned} \tag{4.24}$$

where $|\mathbf{p}_1|, |\mathbf{p}_2| > k_F$, as dictated by Pauli's exclusion principle, and $e_{\mathbf{p}_i}$ denotes the eigenvalue of the operator $\mathcal{H}_k = \mathbf{k}^2/2m + U_k$ belonging to the eigenstate of momentum $\mathbf{p}_i$. It clearly appears that for $W = e_{\mathbf{h}_1} + e_{\mathbf{h}_2}$ the sum in Eq.(4.24) includes the contributions associated with the infinite set of direct *ladder diagrams* involving two hole lines and any numbers of particle lines, which turn out to exhibit the sought-for $A\varrho$ dependence. The ladder diagrams up to third order in v are illustrated in Fig. 4.3.

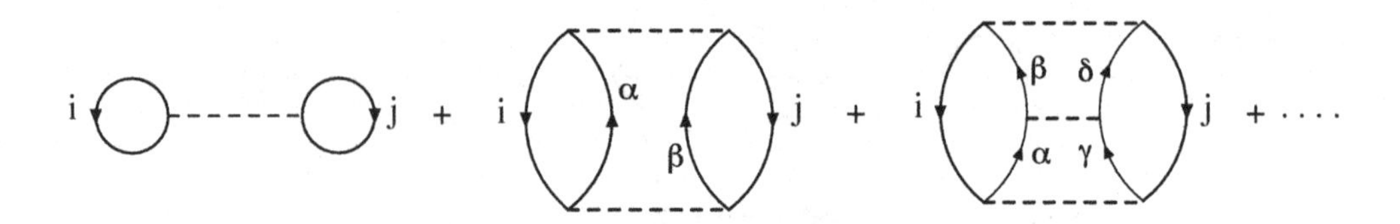

Figure 4.3 Diagrammatic representation of the sum of direct ladder diagrams corresponding to the contributions appearing in the right-hand side of Eq.(4.24). Oriented lines labelled with Latin and Greek letters represent the Green's functions describing nucleons in hole and particle states, respectively.

With the above choice of W, the expression of the ground state energy reduces to

$$E_0 = \sum_{\mathbf{h}_1 \in \{F\}} \frac{\mathbf{h}_1^2}{2m} + \frac{1}{2} \sum_{\mathbf{h}_1 \mathbf{h}_2 \in \{F\}} \{ \langle \mathbf{h}_1, \mathbf{h}_2 | G(e_{\mathbf{h}_1} + e_{\mathbf{h}_2}) | \mathbf{h}_1 \mathbf{h}_2 \rangle - \langle \mathbf{h}_1 \mathbf{h}_2 | G(e_{\mathbf{h}_1} + e_{\mathbf{h}_2}) | \mathbf{h}_2 \mathbf{h}_1 \rangle \} \ . \tag{4.25}$$

The matrix element of the operator G, given by Eq.(4.24), can be conveniently rewritten in terms of the relative and total momenta of the interacting pair

$$\mathbf{k} = \frac{1}{2}(\mathbf{h}_1 - \mathbf{h}_2) \quad , \quad \mathbf{K} = \mathbf{h}_1 + \mathbf{h}_2 \ . \tag{4.26}$$

The resulting expression can be cast in the form

$$\langle \mathbf{k} | G(W) | \mathbf{k} \rangle = \langle \mathbf{k} | v | \mathbf{k} \rangle + \sum_{\mathbf{k}'} f(\mathbf{k}', \mathbf{K}) \frac{\langle \mathbf{k} | v | \mathbf{k}' \rangle \langle \mathbf{k}' | G | \mathbf{k} \rangle}{W - e_{\mathbf{k}'+\mathbf{K}/2} - e_{\mathbf{k}'-\mathbf{K}/2}} \ , \tag{4.27}$$

where we have used the simplified notation

$$\langle \mathbf{k} | G | \mathbf{k} \rangle = \left\langle \mathbf{k} + \frac{1}{2}\mathbf{K}, \mathbf{k} - \frac{1}{2}\mathbf{K} \right| G \left| \mathbf{k} + \frac{1}{2}\mathbf{K}, \mathbf{k} - \frac{1}{2}\mathbf{K} \right\rangle \tag{4.28}$$

$$f(\mathbf{k}', \mathbf{K}) = \theta \left(\left| \mathbf{k}' \pm \frac{\mathbf{K}}{2} \right| - k_F \right) \ . \tag{4.29}$$

The integral equation (4.27) is generally solved by introducing the wave operator Ω, defined as

$$\Omega = 1 + \frac{Q}{W - H_0} v \Omega \ . \tag{4.30}$$

To see how this works, consider that the above equation, implying

$$G = v\Omega \ , \tag{4.31}$$

can be rewritten in coordinate space in the form

$$\psi_{\mathbf{k}}(\mathbf{r}) = \phi_{\mathbf{k}}(\mathbf{r}) + \int \mathcal{G}(\mathbf{r}, \mathbf{r}') v(\mathbf{r}') \psi_{\mathbf{k}}(\mathbf{r}') \ , \tag{4.32}$$

where $\psi_{\mathbf{k}}(\mathbf{r}) = \langle \mathbf{r}|\Omega|\mathbf{k}\rangle$ and $\phi_{\mathbf{k}}(\mathbf{r}) = \langle \mathbf{r}|\mathbf{k}\rangle$ denote the scattering wave function of the two-nucleon system and the corresponding wave function in the absence of interactions, respectively, while the kernel of he integral equation is given by

$$\mathcal{G}(\mathbf{r},\mathbf{r}') = \sum_{\mathbf{k}'} \frac{f(\mathbf{k}',\mathbf{K})}{W - e_{\mathbf{k}'+\mathbf{K}/2} - e_{\mathbf{k}'-\mathbf{K}/2}} \, e^{i\mathbf{k}\cdot(\mathbf{r}-\mathbf{r}')} \ . \tag{4.33}$$

Note that both $\psi_{\mathbf{k}}(\mathbf{r})$ and $\mathcal{G}(\mathbf{r},\mathbf{r}')$ are also functions of W and $\mathbf{K}$. In the above equations, this dependence has been omitted to simplify the notation.

Once Eq.(4.32)—known as Bethe-Goldstone equation—is solved, the matrix elements of the operator G can be obtained from

$$\langle \mathbf{k}|G|\mathbf{k}\rangle = \int d^3r \ e^{i\mathbf{k}\cdot\mathbf{r}} v(\mathbf{r}) \psi_{\mathbf{k}}(\mathbf{r}) \ , \tag{4.34}$$

and the ground state energy can be computed using Eq.(4.25).

It has to be emphasised that, although the BBG approach is based on the replacement of the bare NN interaction with the G-matrix, a perturbative expansion in powers of matrix elements of $G(W)$ would not be convergent. Convergence can only be achieved through a suitable choice of W, allowing to collect the contributions of diagrams with a specific number of independent hole lines. For this reason, the perturbative expansion of BBG theory is referred to as hole-line expansion. It has been shown [63] that a diagram involving n hole lines is proportional to κ^{n-1}, where the quantity κ provides a measure of interaction effects on the two-nucleon wave function. At two-hole-line level

$$\kappa = \varrho \int d^3r \ |\zeta(\mathbf{r})|^2 \ , \tag{4.35}$$

with the defect function $\zeta(\mathbf{r})$ being the difference $\phi_{\mathbf{k}}(\mathbf{r}) - \psi_{\mathbf{k}}(\mathbf{r})$ averaged over the Fermi sea. The shape of the product $r\zeta(\mathbf{r})$ in SNM at $k_F = 1.4$ fm^{-1} is illustrated in Fig. 4.4. It is apparent that, owing to the repulsive core of the NN potential, the wave function describing the relative motion of NN pairs is strongly suppressed at $r \lesssim 1$ fm, where $\zeta(\mathbf{r}) > 0$. Typical values of κ, obtained using realistic NN potentials, are in the range $0.1 - 0.2$ [64].

The value of κ, and therefore the convergence of the hole-line expansion, turn out to be strongly affected by the choice of the single-particle potential $U_{\mathbf{k}}$. Many authors employ a generalisation of the self-consistent Hartee-Fock approximation for hole states, which amounts to setting

$$U_{\mathbf{k}} = \begin{cases} \sum_{\mathbf{k}'\in\{F\}} \left[\langle \mathbf{k}\mathbf{k}'|G|\mathbf{k}\mathbf{k}'\rangle - \langle \mathbf{k}\mathbf{k}'|G|\mathbf{k}'\mathbf{k}\rangle\right] & , \ |\mathbf{k}| < k_F \\ 0 & , \ |\mathbf{k}| > k_F \end{cases} \ . \tag{4.36}$$

The above definition, while leading to the cancellation of a large set of diagrams—therefore improving the convergence of the perturbative expansion—exhibits a somewhat unphysical gap at the Fermi surface.

An alternative scheme, in which the Hartee-Fock definition of the potential is extended to momenta larger than the Fermi momentum, in such a way as to make $U_{\mathbf{k}}$ continuous across the Fermi surface, has been first advocated by the Liège group of C. Mahaux and his collaborators [61].

The two-hole-line approximation of BBG theory with $U_{\mathbf{k}}$ defined according to the Hartee-Fock prescription is referred to as Brueckner-Hartee-Fock (BHF) approximation.

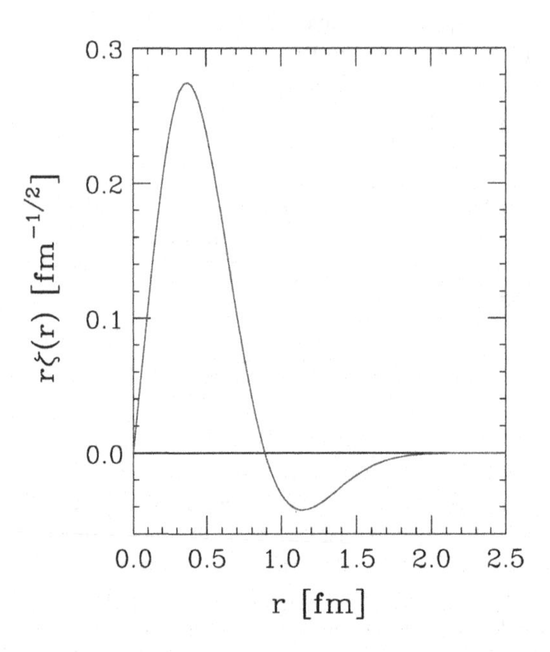

Figure 4.4 Radial dependence of the product $r\zeta(\mathbf{r})$, $\zeta(\mathbf{r})$ being the defect function of SNM at equilibrium density, obtained using the homework potential v_1, see Section4.4.3.

The results of calculations of any nuclear matter properties should obviously be independent of the choice of the single-particle potential. Therefore, the difference between results obtained adopting different prescriptions for $U_{\mathbf{k}}$ provides a measure of the convergence of the hole-line expansion, or lack thereof. This issue has been thoroughly investigated in a number of studies of the three-hole-line correction to the ground-state energy of Eq.(4.25), performed using both the discontinuous and continuous potentials [65, 66].

Figure 4.5 shows the results of calculations carried out within the BBG approach using a nuclear Hamiltonian comprising the AV18 model of the NN potential only [13], without the addition of a NNN potential. The emerging pattern indicates that, in order to bring the energies obtained from the discontinuous and continuous $U_{\mathbf{k}}$ into agreement, the inclusion of sizeable corrections to the two-hole-line approximation—the calculation of which involves non trivial difficulties [65, 67]—are needed at all densities[1]. The continuous choice appears to lead to a better convergence, the three-hole-lines contributions being generally rather small. Note that the range of Fermi momentum spanned by the results of Fig. 4.5 corresponds to densities such that $0.4 \lesssim \varrho/\varrho_0 \lesssim 4.5$.

The results displayed in Fig. 4.5 also illustrate the failure to reproduce the empirical equilibrium properties of SNM using the BHF approximation and a nuclear Hamiltonian including only the two-nucleon potential. This is a long-known problem—independent of the potential model employed to carry out the calculation—first pointed out by F. Coester *et al.* in the early days of nuclear matter theory [68]. The BHF ground-sate energy per particle

[1]We remind the reader that in SNM $k_F = (3\pi^2\varrho/2)^{1/3}$, and the equilibrium density, $\varrho_0 = 0.16$ fm^{-3}, corresponds to $k_F = 1.33$ fm^{-1}.

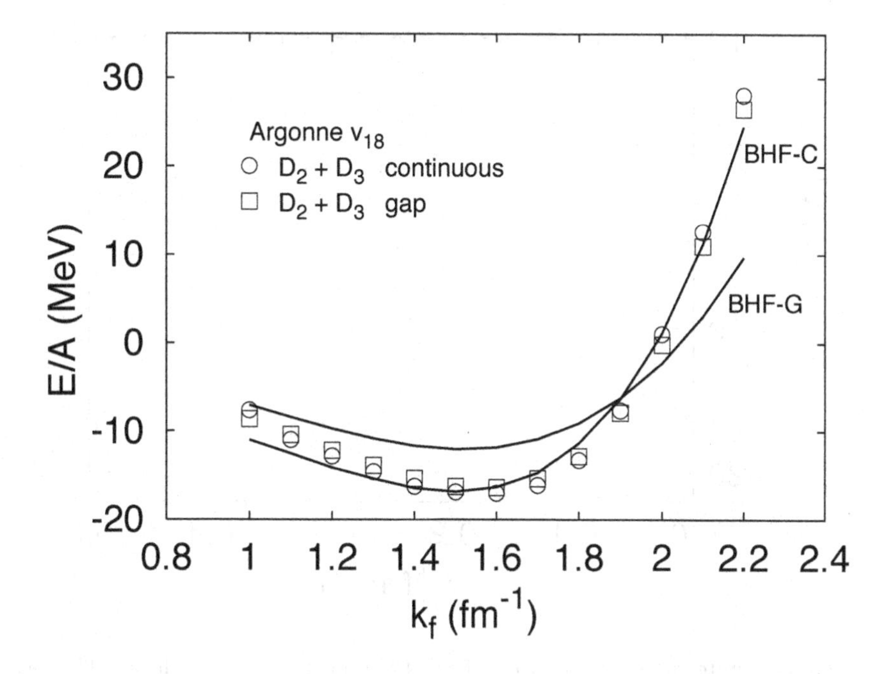

Figure 4.5 Ground-state energy per particle of SNM, obtained from BBG theory using the AV18 potential and different approximation schemes. The full lines labelled BHF-G and BHF-C represent the results of calculations performed within the BHF approximation using the discontinuous and continuous potential, respectively. Squares and circles represent the corresponding results obtained including three-hole-line contributions [66].

exhibits a minimum as a function of the Fermi momentum k_F, but its position lies far beyond the value inferred from the analysis of nuclear densities, $k_F = 1.33 \text{ fm}^{-1}$.

Irreducible three-nucleon interactions, discussed in Chapter 2, are known to be critical in determining the density dependence of the ground-state energy at and beyond the equilibrium point. Their effect has been thoroughly analysed within the BBG approach by X. Zohu *et al.* [69]. The results of this study are summarised in Fig. 4.6, showing a comparison between the energy per nucleon of SNM obtained from the AV18 NN potential, with and without inclusion of the phenomenological UIX NNN potential [20]. Note that, in this case, the parameters involved in Eq.(2.15) have been determined in such a way to reproduce the empirical values of both the energy per particle and the density. It clearly appears that, besides producing a shift of the equilibrium point to lower density, three-nucleon forces provide a large repulsive contribution, that significantly affects the slope of E_0/A—determining the matter pressure—at $\varrho \gg \varrho_0$.

It has to be mentioned that within the BBG method the three-nucleon potential is reduced to a density-dependent two-body potential, averaged over the position and the spin-isospin quantum numbers of the third interacting particle [70].

4.4 THE JASTROW VARIATIONAL APPROACH

The variational approach to the many-body problem with strong forces—widely employed to study the properties of a variety of systems, ranging from the Bose and Fermi hard-

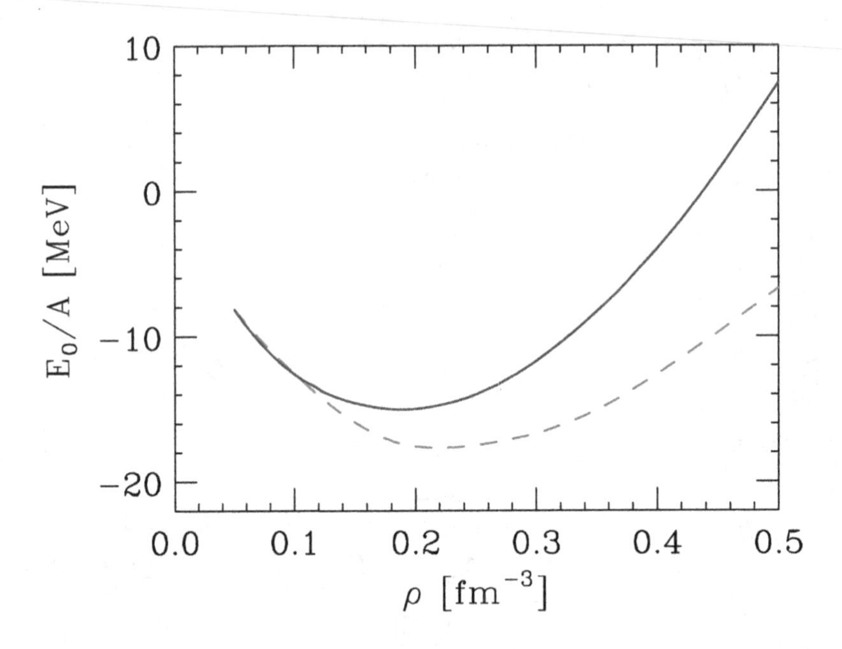

Figure 4.6 Ground-state energy per particle of SNM obtained from BBG theory. The dashed line shows the results of a calculation carried out within the BHF approximation using the AV18 NN potential only, while the solid line represents the results obtained including the contribution of three-nucleon interactions, described by the UIX potential [69].

sphere gases to liquid helium and nuclear matter—was first proposed by R. Jastrow in the 1950s [71]. Within this scheme, the normalised trial ground-state wave function of the Hamiltonian (2.1) is written in the form

$$|\Psi_T\rangle \equiv \frac{F|\Phi_0\rangle}{\langle\Phi_0|F^\dagger F|\Phi_0\rangle^{1/2}} , \tag{4.37}$$

where $|\Phi_0\rangle$ denotes the Fermi gas ground state.

The A-body operator F, describing the effects of correlations among nucleons, is written as a product of two-body operators, whose structure mirrors the form of the potential described in Chapter 2. The resulting expression, to be compared with Eq.(2.12), is

$$F \equiv \mathcal{S} \prod_{i<j} F_{ij} , \tag{4.38}$$

with

$$F_{ij} = \sum_p f^p(r_{ij}) O^p_{ij} . \tag{4.39}$$

Note that the symmetrization operator $\mathcal{S}$ is needed to fulfill the requirement of antisymmetry of $|\Psi_T\rangle$ under particle exchange, since $|\Phi_0\rangle$ is antisymmetric but, in general, $[O^p_{ij}, O^q_{jk}] \neq 0$.

The radial dependence of the correlation functions $f^p(r_{ij})$ is determined by minimizing the expectation value of the Hamiltonian in the correlated ground state

$$E_V = \langle\Psi_T|H|\Psi_T\rangle \geq E_0 , \tag{4.40}$$

that is, by imposing the conditions

$$\frac{\delta E_V}{\delta f^p} = 0 \ . \tag{4.41}$$

The short-distance behaviour of the correlation functions is largely shaped by the strongly repulsive core of the NN potential, producing a drastic suppression of the probability to find two nucleons at relative distance $r_{ij} \lesssim 1$ fm, as illustrated in Fig. 4.7. In the lower panel, the typical probability distributions corresponding to the correlated wave function of two interacting nucleons in a state of total spin and isospin $S = 0$ and $T = 1$ and vanishing angular momentum[2],

$$\psi_{\mathbf{k}}(r) = F_{10}(r)\phi_{\mathbf{k}}(r) \quad , \quad \phi_{\mathbf{k}}(r) \propto \frac{\sin kr}{kr} \ , \tag{4.42}$$

is compared to the Fermi gas probability distribution $r^2|\phi_{\mathbf{k}}(r)|^2$. The relative momentum is set to the average value $k = \sqrt{3/5}\ k_F$, with $k_F = 1.33$ fm^{-1}. The screening effect of correlations is clearly visible, and its range appears to closely reflect the radial dependence of the NN potential, displayed in the upper panel.

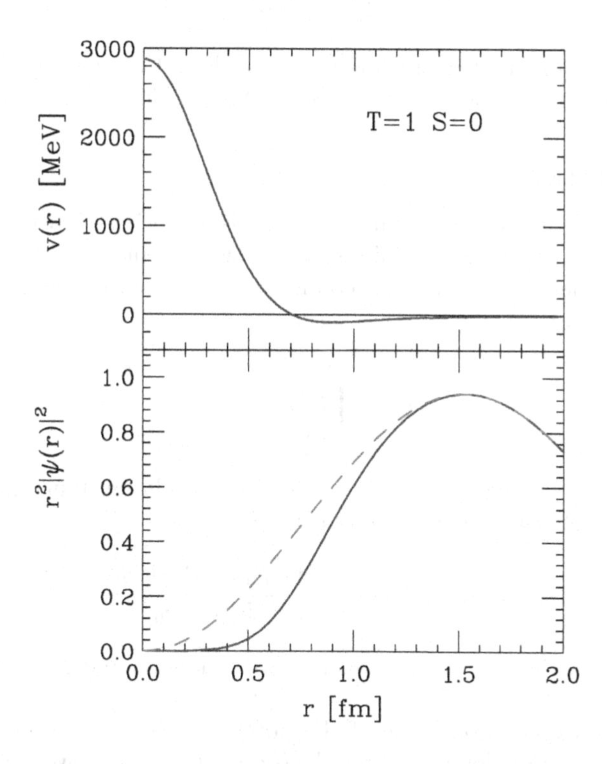

Figure 4.7 The solid and dashed lines in the lower panel represent the probability distributions associated with the correlated and Fermi gas wave functions of two-nucleons having $S = 0$, $T = 1$, vanishing angular momentum, and $k = \sqrt{3/5}\ k_F$, with $k_F = 1.33$ fm^{-1}. For comparison, the upper panel shows the central component of the AV6P potential in the same channel [14].

[2]The projection of the correlation function F_{ij} into states of given S and T, F_{TS}, can be easily obtained from linear combinations of the f^p with $p = 1, \ldots, 6$.

In principle, the variational approach based on a correlated trial wave function, embodying all non perturbative effects of NN interactions, allows to obtain accurate upper bounds to the ground-state energy of nuclear matter. However, in practice the calculation of the expectation value (4.40) with the wave function defined by Eqs.(4.37)-(4.39) involves irreducible $3A$-dimensional integrations whose complexity increases rapidly with A, and becomes intractable already at the level of nuclei such as carbon or oxygen.

The formalism designed to overcome this problem, reminiscent of the cluster expansion techniques developed to calculate the partition function of classical liquids [72], is based on the observation that, owing to the short range nature of correlations, the A-body operator F of Eq.(4.38) exhibits the cluster decomposition property, implying that if any subset of m nucleons, labeled by the indices $i_1 \ldots i_m$, is moved away from the rest, labelled by indices $i_{m+1} \ldots \mathrm{A}$, F reduces to the factorised form

$$F(1 \ldots \mathrm{A}) = F_m(i_1 \ldots i_m) F_{\mathrm{A}-m}(i_{m+1} \ldots i_{\mathrm{A}}) \ . \tag{4.43}$$

From the above result it follows that, because $F_1(i) \equiv 1$ and $F_2(ij) \equiv F_{ij}$, the two-nucleon correlation function satisfies the asymptotic condition

$$\lim_{r_{ij} \to \infty} F_{ij} = 1 \ . \tag{4.44}$$

The above property is apparent in the lower panel of Fig. 4.7, showing that the correlated and uncorrelated wave functions, $\psi_{\mathbf{k}}(r)$ and $\phi_{\mathbf{k}}(r)$, become identical in the limit of large r.

4.4.1 Cluster expansion

The basic elements of the cluster expansion formalism are best illustrated considering a simplified—but still fairly realistic—spin-isospin independent and spherically symmetric potential, $v(r_{ij})$, and neglecting three-nucleon forces altogether. In this case, the correlation operator of Eq.(4.39) reduces to

$$F = \prod_{j>i=1}^{A} f(r_{ij}) \ , \tag{4.45}$$

with the function $f(r_{ij})$ being referred to as Jastrow correlation function, and the Hamiltonian expectation value (4.40) can be rewritten in the form

$$E_V = \langle \Psi_T | H | \Psi_T \rangle = \frac{\langle \Phi_0 | F(T+V) F | \Phi_0 \rangle}{\langle \Phi_0 | F^2 | \Phi_0 \rangle} \ , \tag{4.46}$$

where T and V denote the operators associated withe kinetic and potential energy contributions.

The ground-state expectation value of the potential can be conveniently expressed in terms of the distribution function describing the probability to find two particles at positions $\mathbf{r}_1$ and $\mathbf{r}_2$, defined as

$$p^{(2)}(\mathbf{r}_1, \mathbf{r}_2) = \mathrm{A}(\mathrm{A}-1) \frac{\mathrm{Tr}_1 \mathrm{Tr}_2 \int dx_3 \ldots dx_{\mathrm{A}} |\Psi_T(x_1 \ldots x_{\mathrm{A}})|^2}{\int dx_1 \ldots dx_{\mathrm{A}} |\Psi_T(x_1 \ldots x_{\mathrm{A}})|^2} = \frac{\mathcal{N}}{\mathcal{D}} \ . \tag{4.47}$$

Here, the symbol $x_i \equiv (\mathbf{r}_i, \sigma_i, \tau_i)$ collectively specifies position and discrete quantum numbers of the i-th nucleon, $\mathrm{Tr_i}$ denotes a trace in spin-isospin space, and $dx_i = \mathrm{Tr}_i \ d^3 r_i$.

In translationally invariant systems, $p^{(2)}(\mathbf{r}_1, \mathbf{r}_2)$ only depends on the difference $r = |\mathbf{r}_1 - \mathbf{r}_2|$, and can be written in the form

$$p^{(2)}(\mathbf{r}_1, \mathbf{r}_2) = \varrho^2 g_2(r) , \tag{4.48}$$

where $g_2(r)$, dubbed pair distribution function, exhibits the asymptotic behaviour

$$\lim_{r\to\infty} g_2(r) = 1 + O(A^{-1}) , \tag{4.49}$$

and satisfies the normalisation condition

$$\varrho \int d^3r [g_2(r) - 1] = -1 . \tag{4.50}$$

The above equation can be readily obtained from the requirement of conservation of probability, implying

$$p^{(1)}(\mathbf{r}_1) = \frac{1}{A-1} \int d^3r_2 \; p^{(2)}(\mathbf{r}_1, \mathbf{r}_2) , \tag{4.51}$$

using Eq.(4.48).

Owing to the symmetry of the wave function Ψ_T, the expectation value of the potential—a divergent quantity in the thermodynamic limit—reduces to

$$\langle \Psi_T | V | \Psi_T \rangle = \langle \Psi_T | \sum_{j>i=1}^{A} v(r_{ij}) \; | \Psi_T \rangle = \frac{A(A-1)}{2} \langle \Psi_T | v(r_{12}) | \Psi_T \rangle , \tag{4.52}$$

and dividing by the number of particles we finally obtain the finite result

$$\frac{1}{A} \langle \Psi_T | V | \Psi_T \rangle = \frac{1}{A} \frac{\langle \Phi_0 | F \; \sum_{j>i=1}^{A} v(r_{ij}) \; F | \Phi_0 \rangle}{\langle \Phi_0 | F F | \Phi_0 \rangle} = \frac{\varrho}{2} \int d^3r \; v(r_{12}) g_2(r_{12}) . \tag{4.53}$$

We will now describe the calculation of the pair distribution function $g_2(r_{12})$ appearing in the right-hand side of the above equation. The treatment of the kinetic energy contribution to E_V involves additional issues, and will be analysed at a later stage.

The concepts underlying the cluster expansion formalism have been implemented in a variety of different schemes [56, 73, 74, 55, 75]. Our discussion will follow the approach originally developed by S. Fantoni and S. Rosati,in the 1970s [73, 74].

The starting point is the cluster decomposition of the squared A-body correlation function

$$F^2 = \prod_{j>i=1}^{\mathrm{A}} f^2(r_{ij}) = \prod_{j>i=1}^{\mathrm{A}} [1 + h(r_{ij})] , \tag{4.54}$$

where

$$h(r_{ij}) = f^2(r_{ij}) - 1 . \tag{4.55}$$

From Eq.(4.44), it follows that $h(r)$ is a short-ranged function vanishing at large r. The expansions of the numerator and denominator appearing in Eq.(4.47), denoted $\mathcal{N}$ and $\mathcal{D}$, will be discussed separately.

Diagrammatic expansion of the numerator

Factorisation of the correlation function involving particles 1 and 2—referred to as active particles—in Eq.(4.54) yields

$$F^2 = f^2(r_{12}) \prod_{i<j\neq 1,2} [1+h(r_{ij})] = f^2(r_{12}) \Bigg[1 + \sum_{i\neq 1,2} X^{(3)}(\mathbf{r_1},\mathbf{r_2};\mathbf{r_i}) + \sum_{i<j\neq 1,2} X^{(4)}(\mathbf{r_1},\mathbf{r_2};\mathbf{r_i},\mathbf{r_j}) + \dots \Bigg] \,, \tag{4.56}$$

where

$$X^{(3)}(\mathbf{r}_1,\mathbf{r}_2;\mathbf{r}_i) = h(r_{1i}) + h(r_{2i}) + h(r_{1i})h(r_{2i}) \,, \tag{4.57}$$

and

$$X^{(4)}(\mathbf{r}_1,\mathbf{r}_2;\mathbf{r}_i,\mathbf{r}_j) = h(r_{ij}) + h(r_{1i})h(r_{2j}) + h(r_{1i})h(r_{1j}) + +h(r_{1i})h(r_{ij}) + h(r_{2i})h(r_{2j}) + h(r_{2i})h(r_{ij}) + h(r_{1i})h(r_{2j})h(r_{ij}) + \dots \,. \tag{4.58}$$

The generic term appearing in the right-hand side of Eq.(4.54), $f^2(r_{12})X^{(n)}$, is a n-body operator taking into account all correlations between the two active particles and n background particles belonging to the nuclear medium.

Substitution of Eq.(4.56) in the numerator of the right-hand side of Eq.(4.47) leads to the expression

$$\mathcal{N} = \mathrm{A}(\mathrm{A}-1)\ \mathrm{Tr}_1\mathrm{Tr}_2\ f^2(r_{12}) \times \int dx_3 \dots dx_\mathrm{A} \Bigg[1 + \sum_{i\neq 1,2} X^{(3)}(\mathbf{r}_1,\mathbf{r}_2,\mathbf{r}_i) + \dots \Bigg] |\Phi_0(x_1\dots x_\mathrm{A})|^2, \tag{4.59}$$

showing that the integration involving the n-body contribution $X^{(n)}$ can be written in terms of the n-body distribution function associated with the Fermi gas ground-state wave function Φ_0, defined as

$$g_n^{FG}(\mathbf{r}_1,\dots,\mathbf{r}_n) = \frac{1}{\varrho^n}\frac{\mathrm{A}!}{(\mathrm{A}-n)!}\frac{\mathrm{Tr}_1\dots\mathrm{Tr}_n \int dx_{n+1}\dots dx_\mathrm{A}|\Phi_0(x_1\dots x_\mathrm{A})|^2}{\int dx_1\dots dx_\mathrm{A}|\Phi_0(x_1\dots x_\mathrm{A})|^2} \,. \tag{4.60}$$

The resulting expression is

$$\mathcal{N} = f^2(r_{12}) \Bigg[g_s^{FG}(r_{12}) + \sum_{n=3}^{A} \frac{\varrho^{(n-2)}}{(n-2)!} \int dx_3\dots dx_\mathrm{A}\ X^{(n)}(\mathbf{r}_1,\mathbf{r}_2;\mathbf{r}_3\dots\mathbf{r}_n) g_n^{FG}(\mathbf{r}_1,\mathbf{r}_2,\mathbf{r}_3\dots\mathbf{r}_n) \Bigg] \,. \tag{4.61}$$

The integrals over $x_{n+1},\dots,x_\mathrm{A}$ can be carried out exploiting the properties of determinants and orthogonality of the single-nucleon states $\phi_\mathbf{k}$ of Eq.(3.1). As a result, we obtain

$$g_n^{FG}(\mathbf{r}_1,\dots,\mathbf{r}_n) = \frac{1}{\varrho^n} \sum_{\mathbf{k}_1<\dots<\mathbf{k}_n} \mathrm{Tr}_1\dots\mathrm{Tr}_n \Big\{ \phi^\dagger_{\mathbf{k}_1}(x_1)\dots\phi^\dagger_{\mathbf{k}_n}(x_n)\mathcal{A}\big[\phi_{\mathbf{k}_1}(x_1)\dots\phi_{\mathbf{k}_n}(x_n)\big] \Big\} \,, \tag{4.62}$$

where the operator $\mathcal{A}$ antisymmetrises the product on its right, and all momenta belong to the Fermi sea. Note that the above equation states the obvious result

$$g^{FG}_{n>A}(\mathbf{r}_1, \ldots, \mathbf{r}_n) = 0 \ . \tag{4.63}$$

For $n \leq A$, the explicit expression of the $g^{FG}_n(\mathbf{r}_1, \ldots, \mathbf{r}_n)$ can be written in terms of the density matrix of the non interacting Fermi gas, ϱ_{ij}, defined as

$$\varrho_{ij} = \varrho \ \ell_{ij} = \nu \sum_{|\mathbf{k}|<k_F} \frac{1}{V} e^{i\mathbf{k}\cdot(\mathbf{r}_i - \mathbf{r}_j)} = \varrho \ \ell(k_F r_{ij}) \ , \tag{4.64}$$

with the Slater function given by

$$\ell(x) = 3 \ \frac{\sin x - x \cos x}{x^3} \ , \tag{4.65}$$

implying $\ell_{ii} = 1$, and $\varrho_{ii} = \varrho$. For example, for $n = 2$ and 3, one finds

$$g^{FG}_2(r_{12}) = 1 - \frac{1}{\nu} \ell^2(k_F r_{12}) \ , \tag{4.66}$$

and

$$\begin{aligned} g^{FG}_3(r_{12}, r_{13}, r_{23}) = 1 &- \frac{1}{\nu}\ell^2(k_F r_{12}) - \frac{1}{\nu}\ell^2(k_F r_{13}) - \frac{1}{\nu}\ell^2(k_F r_{23}) \\ &+ \frac{2}{\nu^2}\ell(k_F r_{12})\ell(k_F r_{13})\ell(k_F r_{23}) \ . \end{aligned} \tag{4.67}$$

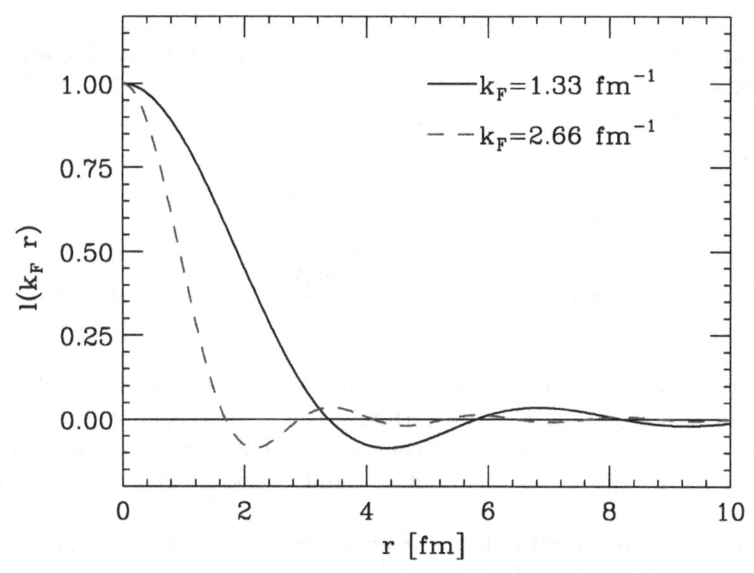

Figure 4.8 Radial and Fermi momentum dependence of the Slater function defined by Eq.(4.65). The solid and dashed lines correspond to $k_F = 1.33$ and $2.66 \ \text{fm}^{-1}$, respectively .

Note that the Slater function $\ell(k_F r)$ introduces a correlation between nucleons separated by a distance r, and exhibits the same asymptotic behaviour as $h(r)$, that is

$$\lim_{r\to\infty} \ell(k_F r) = 0 \ . \tag{4.68}$$

However, while the origin of the correlations described by $f(r)$ is dynamical, those described by $\ell(k_F r)$ arise from Fermi statistics and the antisymmetry of the Fermi gas ground state. The shape of $\ell(k_F r)$ for Fermi momenta $k_F = 1.33$ and 2.66 fm^{-1} is displayed in Fig. 4.8. The range of statistical correlations turns out to decrease with increasing Fermi momentum, and becomes vanishingly small in the limit $k_F \to \infty$, corresponding to infinite density.

The terms appearing in the cluster expansion of $\mathcal{N}$ can be conveniently represented by diagrams, that can be seen a generalisation of the Ursel-Mayer graphs employed in the theory of classical liquids [72]. The diagrammatic rules are the following:

- Each diagram comprises dots, or vertices, associated with the coordinates $\mathbf{r}_i$ and connected by both dynamical and statistical correlation lines. Open dots, also referred to as external points, represent the active particles, labelled 1 and 2, while filled dots, or internal points, correspond to passive, or background, particles. Integration over the coordinates of a passive particle leads to the appearance of a factor ϱ.
- Dashed lines, representing dynamical correlations described by the functions $h(\mathbf{r}_{ij})$ of Eq.(4.55), cannot be superimposed.
- Statistical correlations described by the function $-\ell(k_F r_{ij})/\nu$, with ℓ given by Eq.(4.65), are represented by oriented solid lines, dubbed exchange lines. Exchange lines form closed loops, that can not have common vertices. Loops of oriented lines involving $n > 2$ vertices are multiplied by a factor -2ν. Loops involving only two vertices contribute a factor $-\ell^2(k_F r_{ij})/\nu$.
- Vertices corresponding to passive particles must be reached by at least one correlation line.

As an example, Fig. 4.9 shows two diagrams, corresponding to the following terms of the cluster expansion[3]

$$(4.9.a) = \varrho \int d\mathbf{r}_3 h(r_{13}) h(r_{23}) \,, \tag{4.69}$$

$$(4.9.b) = (-2\nu)\Big(-\frac{\ell(r_{12})}{\nu}\Big) \times \varrho^3 \int d\mathbf{r}_3 d\mathbf{r}_4 d\mathbf{r}_5 \Big(-\frac{\ell(r_{23})}{\nu}\Big)\Big(-\frac{\ell(r_{13})}{\nu}\Big)\Big(-\frac{\ell^2(r_{45})}{\nu}\Big) h(r_{13}) h(r_{45}) \,. \tag{4.70}$$

For comparison, two diagrams forbidden by the above rules are displayed in Fig. 4.10. Diagram (4.10.a) is not allowed because vertex 3, corresponding to a passive particle, is not reached by any correlation lines, and there are two superimposed dashed lines joining vertices 2 and 4. Diagram(4.10.b), on the other hand, exhibits two exchange loops having a common vertex.

The prefactors associated with cluster contributions only include a factor ϱ for each integration over the coordinates of a passive particle, and the factor -2ν associated with exchange loops involving more than two particles. The factor $1/(n-2)!$ in Eq. (4.61), accounting for the number of permutations of the $n-2$ indices associated with passive particles, is taken care of by considering only topologically different graphs, that is, by treating the labels of the filled dots as dummy indices. The only remnant of that factor is a coefficient s^{-1}, where the symmetry number s is the count of permutations of the

[3] To simplify the notation, in the following we denote $\ell(k_F r) = \ell(r)$.

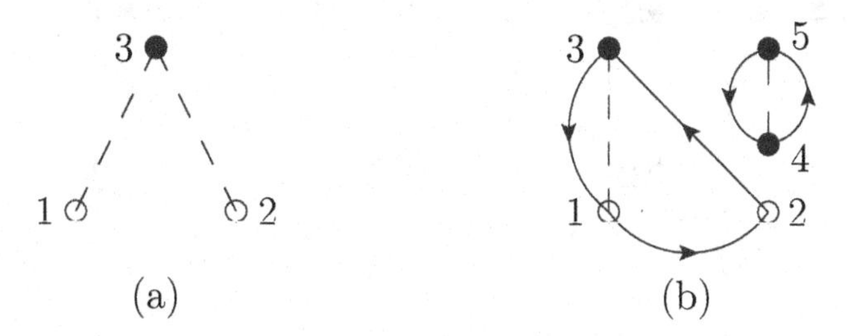

Figure 4.9 Examples of diagrams representing contributions to the cluster expansion of the pair distribution function.

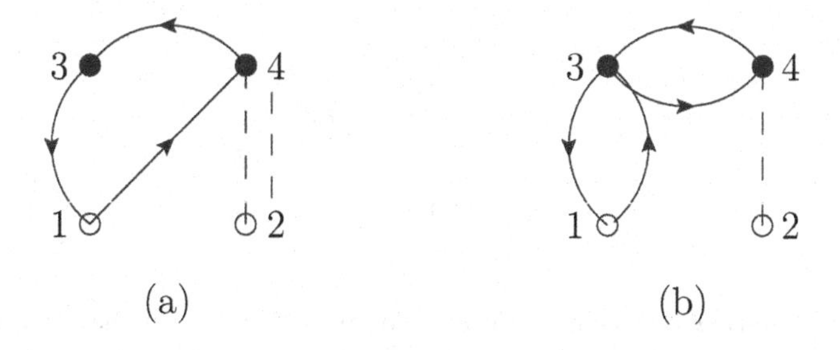

Figure 4.10 Examples of non-allowed cluster diagrams.

labels associated with the filled dots that leave the cluster term unchanged. As an example, diagrams (a) and (b) of Fig. 4.11 are topologically identical, and give identical contributions

$$(4.11.a) = \varrho^2 \int d\mathbf{r}_3 d\mathbf{r}_4 h(r_{13}) h(r_{34}) h(r_{24}) \,, \tag{4.71}$$

$$(4.11.b) = \varrho^2 \int d\mathbf{r}_3 d\mathbf{r}_4 h(r_{14}) h(r_{34}) h(r_{23}) \,. \tag{4.72}$$

As a consequence, only one of them must be taken into account, and no prefactor is needed. On the other hand, a prefactor $s^{-1} = 1/2$ is associated with the diagram (a) of Fig. (4.12), because the corresponding cluster term

$$(4.12.a) = \frac{\varrho^2}{2} \int d\mathbf{r}_3 d\mathbf{r}_4 h(r_{13}) h(r_{14}) h(r_{34}) h(r_{23}) h(r_{24}) \,, \tag{4.73}$$

transform into itself under the exchange of $3 \rightleftharpoons 4$.

Diagrams such that each pair of vertices can be connected by a sequence of lines, representing either dynamical or statistical correlations, are referred to as linked diagrams. Linked diagrams corresponding to cluster contributions in which the integrations can not be factorised are called irreducible. For instance, diagrams (4.9.a), (4.11.a),(4.11.b), and

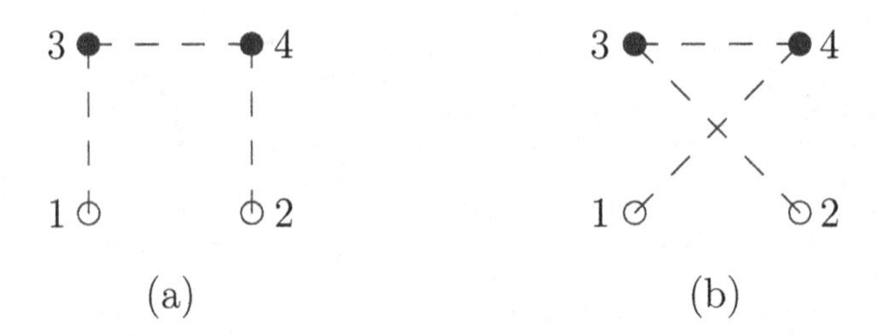

Figure 4.11 Topologically indistinguishable diagrams, yielding identical cluster contributions.

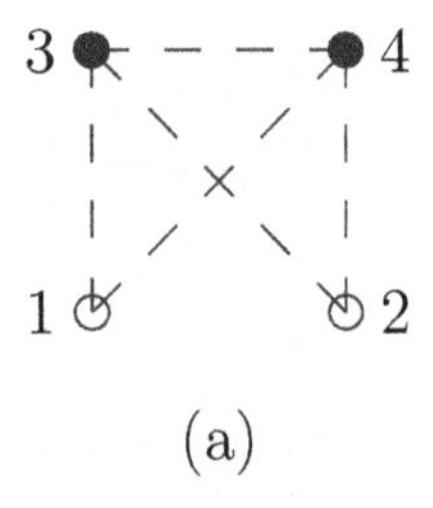

Figure 4.12 Graph requiring the inclusion of a prefactor $s^{-1} = 1/2$ in the corresponding cluster term.

(4.12.a) are linked and irreducible, while diagram (4.9.b) is unlinked. By attaching to the irreducible diagram (4.9.a) the correlation line $h(\mathbf{r}_{34})$, we obtain the reducible diagram of Fig. (4.13), with vertex 3 being the reducibility point. The corresponding contribution

$$(4.13.a) = \varrho^2 \int d\mathbf{r}_3 d\mathbf{r}_4 h(r_{13}) h(r_{23}) h(r_{34}) \,, \tag{4.74}$$

$$= \varrho \int d\mathbf{r}_3 h(r_{13}) h(r_{23}) \times \varrho \int d\mathbf{r}_{34} h(r_{34}) \,, \tag{4.75}$$

is nothing but the product between the contribution associated with diagram (4.9.a) and the one corresponding to the additional element.

Translational invariance gives rise to appearance of a factor V for each unlinked part of the diagram, except the one containing the vertices 1 and 2, representing the active particles. Therefore, a cluster diagram is proportional to $V^{n_u - 1}$, n_u being the number of unlinked parts. For example, the contribution os the linked diagram (4.9.a) is proportional to V^0 while that of diagram (4.9.b), having two disconnected parts, is proportional to V.

Diagrammatic expansion of the denominator

The same procedure followed for the expansion of the numerator of Eq.(4.47) can be used for the denominator. In this case there are no active particles, and the squared correlation

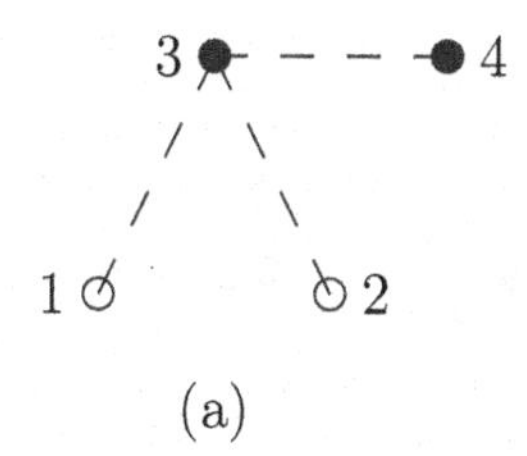

Figure 4.13 Example of linked and reducible diagram.

function F^2 can be conveniently written in the form

$$F^2 = \left[1 + \sum_{i<j} X^{(2)}(\mathbf{r}_i, \mathbf{r}_j) + \sum_{i<j<k} X^{(3)}(\mathbf{r}_i, \mathbf{r}_j, \mathbf{r}_k) + \dots \right] , \tag{4.76}$$

where the term $X^{(n)}$ involves correlations among n particles. The explicit expressions for $n = 2$ and 3 read

$$X^{(2)}(\mathbf{r}_i, \mathbf{r}_j) = h(r_{ij}) , \tag{4.77}$$

and

$$X^{(3)}(\mathbf{r}_i, \mathbf{r}_j, \mathbf{r}_k) = h(r_{ij})h(r_{ik}) + h(r_{ik})h(r_{jk}) + h(r_{ij})h(r_{jk}) + h(r_{ij})h(r_{jk})h(r_{ik}) . \tag{4.78}$$

Substituting the expansion of Eq. (4.76) in the denominator of Eq. (4.47) and exploiting the invariance of $|\Phi_0|^2$ under any two-particle exchanges, one finds

$$\mathcal{D} = 1 + \sum_{n=2}^{A} \frac{\varrho^n}{n!} \int d\mathbf{r}_1 \dots d\mathbf{r}_n \; g_n^{FG}(\mathbf{r}_1, \dots, \mathbf{r}_n) X^{(n)}(\mathbf{r}_1, \dots, \mathbf{r}_n) , \tag{4.79}$$

with g_n^{FG} defined in Eq. (4.62).

A comparison between Eqs. (4.61) and (4.79) suggests that the numerator and the denominator in Eq.(4.47) can be represented using very similar diagrammatic rules. The only difference is that the cluster diagrams of the denominator, few examples of which are given in Fig. 4.14, comprise filled points only.

Pair distribution function

The ratio of Eq.(4.47) involves two infinite series of cluster terms—corresponding to the expansion of the numerator and the denominator—that can be represented by diagrams according to the rules described in the previous sections.

Let us denote by $\mathcal{L}_n$ the contribution of a generic n-body linked diagram, reducible or irreducible, appearing in the expansion of the numerator $\mathcal{N}$, and by $\mathcal{U}_m$ the sum of all m-body contributions represented by diagrams disconnected from $\mathcal{L}_n$.

From the above definitions, it follows that the denominator of Eq.(4.47) can be written in the form

$$\mathcal{D} = \sum_{m=0}^{A} \mathcal{U}_m . \tag{4.80}$$

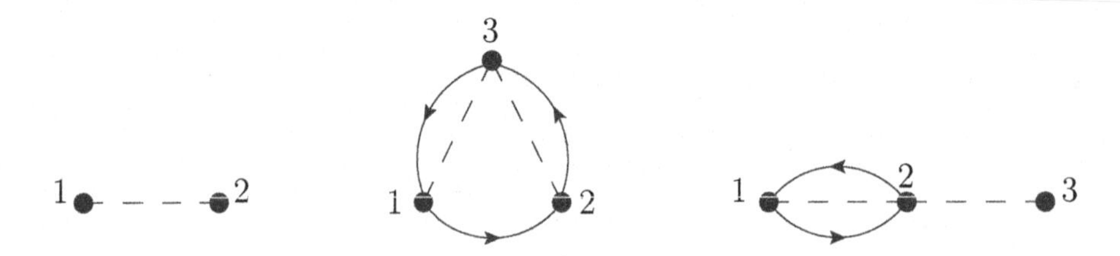

Figure 4.14 Examples of cluster diagrams contributing to the denominator of Eq.(4.47).

In addition, for any given $\mathcal{L}_n$ the cluster expansion of the numerator includes m-body terms obtained from the product $\mathcal{L}_n \times \mathcal{U}_{m-n}$. Therefore, the expression of $\mathcal{N}$ exhibits the form

$$\mathcal{N} = \sum_{n=2}^{\mathrm{A}} \sum_{\mathcal{L}_n} \mathcal{L}_n \times \sum_{m=0}^{\mathrm{A}} \mathcal{U}_m \, . \tag{4.81}$$

The above result is illustrated in Fig. 4.15. In this example, $n = 3$, $\mathcal{L}_3$ corresponds to diagram (4.9.a), and the sum of the $\mathcal{U}_m$ is represented by the diagrams enclosed in square brackets.

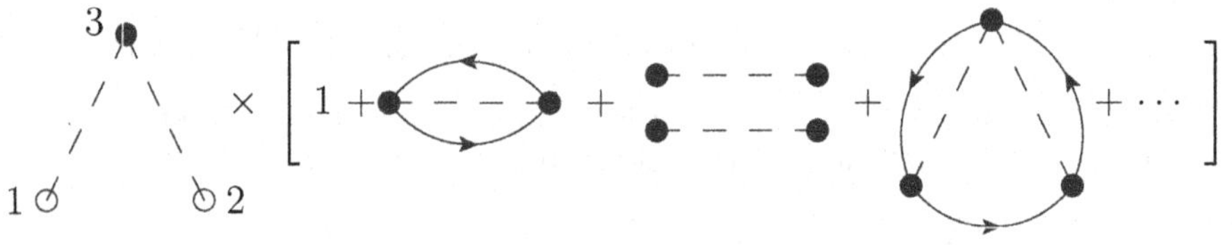

Figure 4.15 Series of unlinked diagrams associated with the linked diagram $\mathcal{L}_3$ given by Fig.(4.9.a).

Equations (4.80) and (4.81) show that in the cluster expansion of the pair distribution function, involving the ratio $\mathcal{N}/\mathcal{D}$, the contributions of unlinked graphs cancel, and the final result reduces to

$$\begin{aligned} g_2(r_{12}) &= \frac{1}{\varrho^2} \sum_{n=2}^{A} \sum_{\mathcal{L}_n} \mathcal{L}_n \\ &= f^2(r_{12}) \Big[g_2^{FG}(r_{12}) + \sum_{n>2} X_L^{(n)}(\mathbf{r}_1, \mathbf{r}_2) \Big] \, , \end{aligned} \tag{4.82}$$

where g_2^{FG} is the Fermi gas distribution function of Eq.(4.66), while X_L^n denotes the sum off all linked contributions to the cluster expansion of $\mathcal{N}$ involving $n-2$ background nucleons. It is worth noting that, owing to Eq.(4.63), the above result, whose derivation is straightforward in the nuclear matter limit $\mathrm{A} \to \infty$, holds true for any finite A as well.

The cancellation of the divergent terms associated with unlinked contributions—leading to finite results in calculations of physical quantities in the thermodynamic limit—is a general properties of diagrammatic expansion techniques, which also applies to the case of spin-isospin dependent correlations.

By exploiting the properties of the Fermi gas density matrix $\ell(r)$ defined by Eq.(4.64), satisfying

$$\int d^3r\ \ell(r) = \int d^3r\ \ell^2(r) = \frac{\nu}{\varrho}\ , \tag{4.83}$$

and

$$\int d^3r_3\ \ell(r_{13})\ell(r_{32}) = \frac{\nu}{\varrho}\ell(r_{12})\ , \tag{4.84}$$

it has been shown that the cluster expansion of translationally invariant Fermi systems only includes the contribution of irreducible diagrams [73]. However, unlike the case of unlinked diagrams, the cancellation of reducible diagrams only occur in the case of spin-isospin independent correlations.

The cluster terms contributing to $g_2(r)$ can be grouped according to different schemes. Collecting all diagrams with the same number of vertices leads to an expansion whose m-th order term, obtained restricting the sum in Eq.(4.82) to $n = m$, turns out to be proportional to ϱ^{m-2}. For example, the contribution of diagram (a) of Fig. 4.9, having $m = 3$, is of order ϱ. The validity of the lowest order—or two-body cluster—approximation derived within this scheme, yielding

$$g_2^{(2)}(r_{12}) = f^2(r_{12})\left[1 - \frac{1}{\nu}\ell^2(k_F r_{12})\right]\ , \tag{4.85}$$

is clearly limited to the low-density regime.

An alternative expansion is obtained by ordering the cluster terms according to the number of dynamical correlations $h(r_{ij})$. A remarkable property of this expansion, whose 0-th order term coincides with the Fermi gas result, is that the resulting radial distribution function satisfies the normalisation condition (4.50) at all orders. On the other hand, at any given order the expansion in powers of h involves diagrams belonging to different orders of the expansion in number of vertices, thus mixing up different density dependences. For example, the term linear in h includes diagrams involving both one and two passive particles, whose contributions are proportional to ϱ and ϱ^2, respectively.

4.4.2 Kinetic energy

The expectation value of the kinetic energy in the correlated ground state is given by

$$\langle\Psi_T|T|\Psi_T\rangle = -\frac{1}{2m}\sum_i\langle\Psi_T|\boldsymbol{\nabla}_i^2|\Psi_T\rangle = -\frac{1}{2m}\mathrm{A}\langle\Psi_T|\boldsymbol{\nabla}_1^2|\Psi_T\rangle\ , \tag{4.86}$$

with

$$-\langle\Psi_T|\boldsymbol{\nabla}_1^2|\Psi_T\rangle = -\frac{\int dx_1\ldots dx_{\mathrm{A}}\Phi_0^\dagger F\ \boldsymbol{\nabla}_1{}^2\ F\Phi_0}{\int dx_1\ldots dx_{\mathrm{A}}\Phi_0^\dagger F^2\Phi_0}\ , \tag{4.87}$$

where the dependence of Φ_0 and F on $x_1\ldots x_{\mathrm{A}}$ is understood, and the symmetry properties of the correlated wave-function have been exploited.

Equation (4.87) can be rewritten in momentum space in the form

$$-\frac{1}{2m}\ \langle\Psi_T|\boldsymbol{\nabla}_1^2|\Psi_T\rangle = \int\frac{d^3k}{(2\pi)^3}\frac{k^2}{2m}\ n(k)\ , \tag{4.88}$$

where the momentum distribution $n(k)$, defined by (4.11), can be expressed in terms of the density matrix, $n(r)$, according to

$$n(k) = \int d^3r \ e^{i\mathbf{k}\cdot\mathbf{r}} n(r) \ , \tag{4.89}$$

with

$$n(|\mathbf{r}_1 - \mathbf{r}_{1'}|) = \mathrm{A} \frac{\mathrm{Tr}_1 \mathrm{Tr}_{1'} \int dx_2 \ldots dx_\mathrm{A} \Psi_T^\dagger(x_1, x_2, \ldots, x_\mathrm{A}) \Psi_T(x_{1'}, x_2, \ldots, x_\mathrm{A})}{\int dx_1 \ldots dx_\mathrm{A} |\Psi_T(x_1, x_2, \ldots, x_\mathrm{A})|^2} \ . \tag{4.90}$$

In order to treat the potential and kinetic energy contributions in a consistent fashion, however, the expectation value (4.87) is evaluated mainly in coordinate space. Numerical studies are based on different expressions of the expectation value (4.87), which, while being equivalent in an exact calculation, yield different results at finite order of the cluster expansion. Therefore, the size of the difference between results obtained from different expressions provides a measure of the converge of the expansion.

Kinetic energy expressions

The right-hand side of Eq.(4.87) can be straightforwardly rewritten to obtain

$$\begin{aligned} T_{PB} = -\frac{1}{2m} \sum_i \langle \Psi_T | \boldsymbol{\nabla}_i^2 | \Psi_T \rangle &= -\frac{1}{2m} \frac{1}{\mathcal{N}(\Psi_T)} \sum_i \int dx_1 \ldots dx_\mathrm{A} \Phi_0^\dagger F \Big[F(\boldsymbol{\nabla}_i^2 \Phi_0) \\ &+ 2(\boldsymbol{\nabla}_i F) \cdot (\boldsymbol{\nabla}_i \Phi_0) + \Phi_0(\boldsymbol{\nabla}_i^2 F) \Big] \ , \end{aligned} \tag{4.91}$$

with the normalisation being defined as

$$\mathcal{N}(\Psi_T) = \int dx_1 \ldots dx_\mathrm{A} |\Psi_T(x_1, x_2, \ldots, x_\mathrm{A})|^2 \ . \tag{4.92}$$

The above expression of the kinetic energy is referred to as Pandharipande-Bethe (PB) form [76]. The first term in square brackets yields the ground-state energy of the non interacting Fermi gas, see Eq.(3.10)

$$T_F = \mathrm{A} \ \frac{3}{5} \frac{k_F^2}{2m} \ . \tag{4.93}$$

From the definition of F, Eq.(4.45), implying

$$\boldsymbol{\nabla}_i F = \sum_{j>i} \frac{\boldsymbol{\nabla}_i f(r_{ij})}{f(r_{ij})} F \ , \tag{4.94}$$

and

$$F \ \boldsymbol{\nabla}_i^2 F = \sum_{j>i} \left[\frac{\boldsymbol{\nabla}_i^2 f(r_{ij})}{f(r_{ij})} - \frac{(\boldsymbol{\nabla}_i f(r_{ij}))^2}{f^2(r_{ij})} \right] F^2 + \sum_{k>j>i} \frac{(\boldsymbol{\nabla}_i f(r_{ij})) \cdot (\boldsymbol{\nabla}_i f(r_{ik}))}{f(r_{ik}) f(r_{ik})} F^2 \ , \tag{4.95}$$

it follows that the calculation of the kinetic energy involves the expectation values of two- and three-body operators, and therefore requires the two- and three-body distribution functions.

The Clark-Westhaus form of the kinetic energy [77] is obtained integrating by parts the last contribution in square brackets in the right-hand side of Eq. (4.91) and using the identity

$$\sum_i [(\nabla_i \Phi_0^\dagger) F (\nabla_i F) \Phi_0 - \Phi_0^\dagger (\nabla_i F) F (\nabla_i \Phi_0)] = 0, \tag{4.96}$$

with the result

$$T_{CW} = -\frac{1}{2m} \frac{1}{\mathcal{N}(\Psi_T)} \sum_i \int dx_1 \dots dx_A \Phi_0^\dagger F \Big[F(\nabla_i^2 \Phi_0) - (\nabla_i F)^2 \Phi_0 \Big] \; . \tag{4.97}$$

Collecting together the two-body contribution to the CW kinetic energy and the interaction potential, one can define an *effective potential*

$$w(r) = \frac{1}{m} \left(\frac{\nabla f(r)}{f(r)} \right)^2 + v(r) \; , \tag{4.98}$$

with the operator ∇ acting on r, whose expectation value can be evaluated using the two-nucleon radial distribution function $g_2(r)$. On the other hand, the calculation of the expectation value of the three-nucleon operator arising from the second term in square brackets in Eq.(4.97) requires the three-nucleon distribution function.

The Jackson-Feenberg form of the kinetic energy [78], whose derivation also involves integrations by parts of the right-hand side of Eq.(4.91), reads

$$T_{JF} = -\frac{1}{4m} \frac{1}{\mathcal{N}(\Psi_T)} \sum_i \int dx_1 \dots dx_A \Big[\Phi_0^\dagger F (\nabla_i^2 F \Phi_0) \tag{4.99}$$

$$- 2(\nabla_i \Phi_0^\dagger F) \cdot (\nabla_i F \Phi_0) + (\nabla_i^2 \Phi_0^\dagger F) F \Phi_0 \Big] \; . \tag{4.100}$$

In this case, the two-nucleon contribution, involving derivatives acting on the argument of the correlation function $f(r_{ij})$ only, turns out to be

$$T_{JF}^{(2)} = -\frac{1}{2m} \sum_{j>i} \int dx_1 \dots dx_A \; \Phi_0^\dagger F (\nabla^2 \ln f(r_{ij})) F \Phi_0 \; . \tag{4.101}$$

The remaining contributions give rise to three-body terms involving the Fermi gas density matrix and its derivative.

The three alternative forms of the kinetic energy, while being in principle equivalent, give different results when implemented through a cluster expansion. For example, the contribution of three-body terms turns out to be smaller in the JF form than in the PB or CW forms. Therefore, the JF kinetic energy is less affected by the approximations involved in the derivation of the three-body distribution function. On the other hand, it turns out to be more affected by the description of the two-nucleon distribution function at short distances.

4.4.3 Low-order variational calculation of nuclear matter energy

The Jastrow variational approach and the cluster expansion formalism have been extensively employed to obtain low-order approximations to the energy of nuclear matter using central spin-isospin independent potentials. However, it became soon apparent that these studies are plagued by a fundamental conceptual problem. The unconstrained minimisation of the expectation value of the Hamiltonian computed at any finite order of the cluster expansion yields a variational energy that fails to meet the requirement of boundedness from below.

The above problem has been circumvented constraining the correlation function in such a way as to satisfy an additional condition, such as, for example, normalisation of the pair distribution function according to Eq.(4.50). This procedure does lead to a fast convergence of the cluster expansion, although the resulting energy is not guaranteed to provide an accurate upper bound to the true ground-state energy.

In order to assess the validity of the both the variational and G-matrix approaches, in the 1970s, H. Bethe proposed to perform systematic comparative studies using two central spin- and isospin-independent potentials, obtained from the analysis of NN scattering phase shift in the spin-singlet and spin-triplet S channels. In the literature, these potentials are referred to as Bethe's homework potentials and denoted v_1 and v_2, respectively.

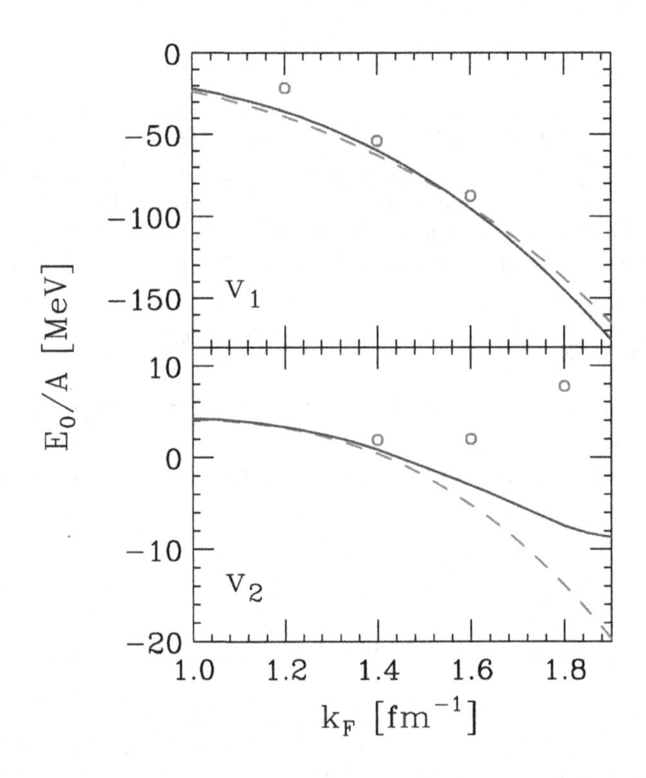

Figure 4.16 Energy per particle of SNM computed using Bethe's homework potentials. The solid and dashed lines show the variational upper bounds obtained taking into account cluster contributions up to two- and three-body level, respectively [79]. The circles represent BHF results obtained with the single nucleon potential $U_{\mathbf{k}}$ of Eq.(4.36) [80].

As an example, Fig. 4.16 shows the energy per nucleon of SNM computed within the variational approach [79]. The results have been obtained including two- and three-body cluster contributions, and using the Jastrow correlation function

$$f(r) = \left[1 - e^{-(r^2/b^2)}\right]^n + g r^m e^{-(r^2/\gamma^2)} \ , \tag{4.102}$$

with the values of the parameters b, n, g, m, and γ determined by energy minimisation with the additional constraint

$$\varrho \int d^3r[1 - f^2(r)]g_2^{FG}(r) = 0 \ . \tag{4.103}$$

Note that the above equation, with g_2^{FG} given by Eq.(4.66), corresponds to Eq.(4.50) in the two-body cluster approximation.

A comparison between the full and dashed lines of Fig. 4.16, corresponding to the two- and three-body cluster approximations, respectively, shows that the lowest-order approximation provides accurate results for the v_1 model over the whole range of Fermi momenta. In the case of the v_2 model, on the other hand, significant corrections arising from three-body cluster contributions appear above $k_F \sim 1.4$ fm^{-1}.

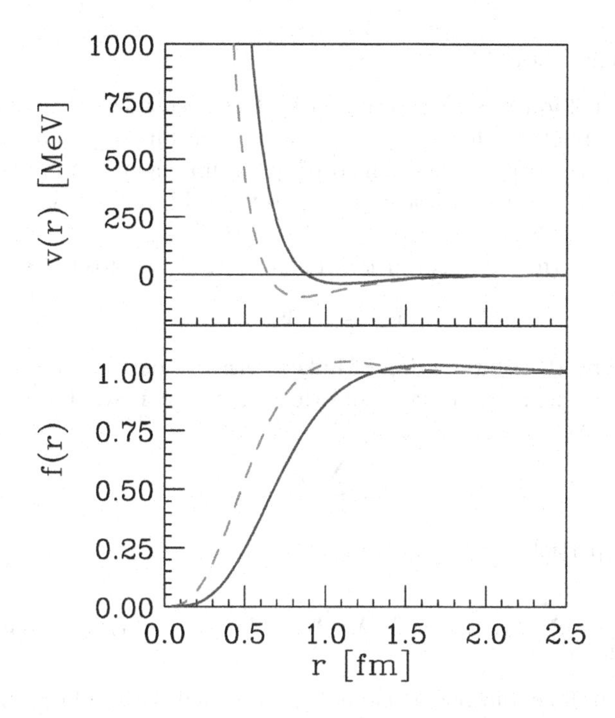

Figure 4.17 The dashed and solid lines of the top panel show the radial dependence of Bethe's homework potentials v_1 and v_2, respectively. The corresponding correlation functions at $k_F = 1.4$ fm^{-1} are displayed in the bottom panel.

The observed pattern can be easily explained considering that the rate of convergence of the cluster expansion can be estimated from the correlation volume

$$\kappa_J = \varrho \int d^3r \ [f(r) - 1]^2 g_2^{FG}(k_F r) \ , \tag{4.104}$$

to be compared to the analogue quantity of BBG theory, Eq.(4.35).

The v_1 and v_2 potentials and the corresponding correlation functions at $k_F = 1.4$ fm^{-1} are displayed in Fig. 4.17. It is apparent that the stronger repulsive core of the v_2 potential produces a stronger suppression of the correlation function, resulting in a larger value of κ_J and a slower convergence of the cluster expansion.

Figure 4.16 also shows results obtained within the BHF approximation [80]. In the case of the v_1 potential, the comparison with the variational energies suggests that the two-hole-line approximation is conceptually similar the two-body cluster approximation. However, the large discrepancies occurring for the v_2 model indicate that the corrections to the lowest order approximation are larger in BBG theory than in the variational approach.

4.5 ADVANCED PERTURBATIVE METHODS

The issues associated with the convergence of the hole-line expansion of BBG theory motivated the development of approaches that, while being similar in spirit to G-matrix perturbation theory, allow to take into account the contribution of higher-order terms in a systematic fashion.

4.5.1 Coupled cluster method

The coupled cluster method was first proposed by F. Coester and H. Kümmel [81] in the 1960s, to describe correlation effects in the nuclear wave function. This approach, which has been also extensively applied in the field of quantum chemistry [82], allows to reduce the description of an interacting A-body system to the solution of a set of coupled n-body equations, with $n = 1, \dots$ A.

The starting point is the exponential *ansatz* for the A-body ground state

$$|\Psi_0\rangle = e^S|\Phi_0\rangle \ , \tag{4.105}$$

which allows to identify the contribution of linked terms in perturbative expansions. Here, $|\Phi_0\rangle$ denotes the Fermi gas ground state, and the operator S is written in the form

$$S = \sum_{n=1}^{A} S_n \ , \tag{4.106}$$

with S_n being the n-particle n-hole excitation operators

$$S_n = \left(\frac{1}{n!}\right)^2 \sum_{\mathbf{h}_1 \dots \mathbf{h}_n \mathbf{p}_1 \dots \mathbf{p}_n} \langle \mathbf{p}_1 \dots \mathbf{p}_n|\chi_n|\mathbf{h}_1 \dots \mathbf{h}_n\rangle a^\dagger_{\mathbf{p}_1} \dots a^\dagger_{\mathbf{p}_n} a_{\mathbf{h}_n} \dots a_{\mathbf{h}_1} \ . \tag{4.107}$$

In the above equation, hole and particle states are labelled with indices $\mathbf{h}_i$ and $\mathbf{p}_i$, respectively, while $a^\dagger_{\mathbf{p}_i}$ and $a_{\mathbf{h}_i}$ denote creation and annihilation operators.

Combining the Schrödinger equation, $H|\Psi_0\rangle = E_0|\Psi_0\rangle$, and Eq.(4.105), one obtains the relation

$$e^{-S}He^S|\Phi_0\rangle = E_0|\Phi_0\rangle \ , \tag{4.108}$$

that can be exploited to derive the expression of the ground-state energy, as well as a set of coupled equations for the S_n. Note that in nuclear matter translational invariance, implying momentum conservation, dictates that $S_1 = 0$.

Multiplication of Eq.(4.108) by $\langle\Phi_0|$ yields the ground state energy

$$E_0 = \langle\Phi_0|e^{-S}He^S|\Phi_0\rangle = \langle\Phi_0|(1 - S_2 - \dots)(T + V)(1 + S_2 + \dots)|\Phi_0\rangle \ , \tag{4.109}$$

while multiplying by the n-particle n-hole state $\langle \mathbf{h}_1 \dots \mathbf{h}_n, \mathbf{p}_1 \dots \mathbf{p}_n|$ leads to a set of A relations involving the S_n

$$\langle \mathbf{h}_1 \dots \mathbf{h}_n, \mathbf{p}_1 \dots \mathbf{p}_n|e^{-S}He^S|\Phi_0\rangle = 0 \quad , \quad n = 2, \dots, \mathrm{A} \ . \tag{4.110}$$

The infinite hierarchy of equations (4.110), referred to as Coupled Cluster (CC) equations, is equivalent to the Scrödinger equation determining the nuclear matter ground-state and the corresponding energy.

Neglecting, for simplicity, three-nucleon interactions Eq.(4.109) can be written in the form

$$E_0 = T_F + \frac{1}{2} \sum_{\mathbf{h}_1\mathbf{h}_2} \langle \mathbf{h}_1\mathbf{h}_2|v|\mathbf{h}_1\mathbf{h}_2\rangle + \frac{1}{4} \sum_{\mathbf{h}_1\mathbf{h}_2\mathbf{p}_1\mathbf{p}_2} \langle \mathbf{h}_1\mathbf{h}_2|v|\mathbf{p}_1\mathbf{p}_2\rangle\langle \mathbf{p}_1\mathbf{p}_2|\chi_2|\mathbf{h}_1\mathbf{h}_2\rangle \tag{4.111}$$

where T_F is the ground-state energy of the Fermi gas and v denotes the NN potential.

In principle, Eq.(4.111) provides a well defined prescription to obtain the ground-state energy. It must be kept in mind, however, that χ_2 is coupled to the higher order amplitudes through Eqs.(4.110). As a consequence, numerical calculations of E_0 necessarily involve the truncation of the hierarchy of CC equations at some finite order m, which amounts to neglecting all χ_n with $n > m$ [83]. Within this scheme, nuclear matter is described as a collection of m-body clusters.

The lowest-order approximation, corresponding to $m = 2$, is reminiscent of the two-hole line approximation of BBG theory. However, besides the particle-particle ladder diagrams contributing to the G-matrix, see Fig. 4.3, it involves a large set of additional terms, such as those corresponding to hole-hole ladder diagrams.

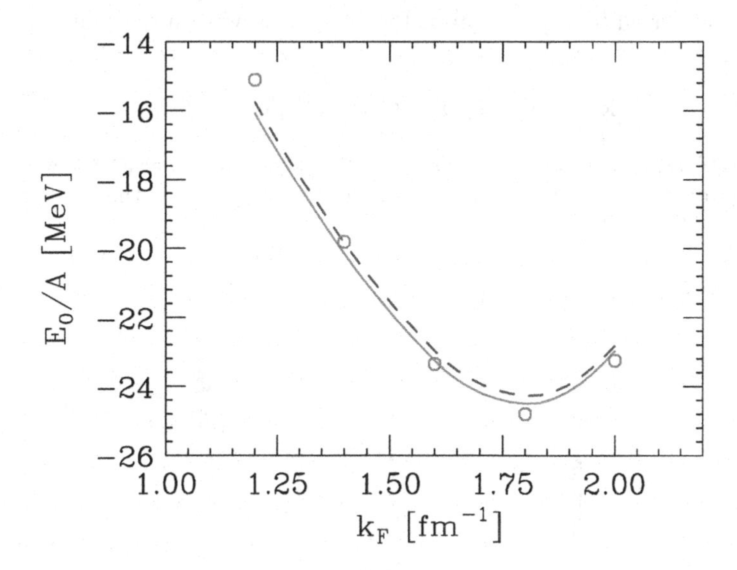

Figure 4.18 Energy per particle of SNM computed using the Coupled Cluster formalism and a N^{3}LO NN potential derived from χEFT. The results represented by the dashed line include the contribution of particle-particle ladders only, while the solid line has been obtained taking into account both particle-particle and hole-hole ladders. For comparison, the open circles show the results obtained using the BHF approximation.

The results of a calculation carried out within the $m = 2$ truncation scheme using a NN potential derived from χEFT at N^{3}LO [84] are displayed in Fig. 4.18 [85]. The energies of

SNM obtained including particle-particle ladder diagrams only or taking also into account hole-hole ladder diagrams—represented by the dashed and solid lines, respectively—are compared with those computed within the BHF approximation, displayed by the open circles. All calculations have been performed adopting the continuous choice for the Hartree-Fock potential, see Sect. 4.3. The results of BBG theory and the Coupled Cluster method turn out to be in fairly good agreement. However, the observed differences do not appear to be ascribable to the contribution of hole-hole ladder diagrams.

The Coupled Cluster approach has been extended to include three-nucleon interactions, although the existing calculations carried out using χEFT Hamiltonian are unable to simultaneously account for the saturation properties of SNM and the binding energies of light nuclei [86]. On the other hand, the approach based on a combination of Coupled Cluster formalism and χEFT dynamics has been shown to provide a fairly accurate description of the properties of a broad range of atomic nuclei [87].

4.5.2 Self-consistent Green's function method

The perturbative scheme employed to perform calculations of the ground-state energy has been generalised to obtain the two-point Green's function introduced in Section 3.3.1, embodying all information on single-nucleon properties in nuclear matter [88]. This approach, referred to as Self-Consistent Green's Function (SCGF) method, has been applied to both atomic nuclei and nuclear matter, using realistic nuclear Hamiltonian comprising two- and three-nucleon potentials. [89].

The SCGF formalism is based on a diagrammatic expansion of the nucleon Green's function $G(\mathbf{k},\omega)$, also referred to as propagator. Figure 4.19 provides a schematic representation of Dyson's equation

$$G(\mathbf{k},\omega) = G_0(\mathbf{k},\omega) + G_0(\mathbf{k},\omega)\Sigma(\mathbf{k},\omega)G(\mathbf{k},\omega) \tag{4.112}$$

in which G_0 and G, represented by the single and double oriented lines, are the free-space and dressed propagator, respectively, while Σ, depicted by the blob, is the irreducible nucleon self energy.

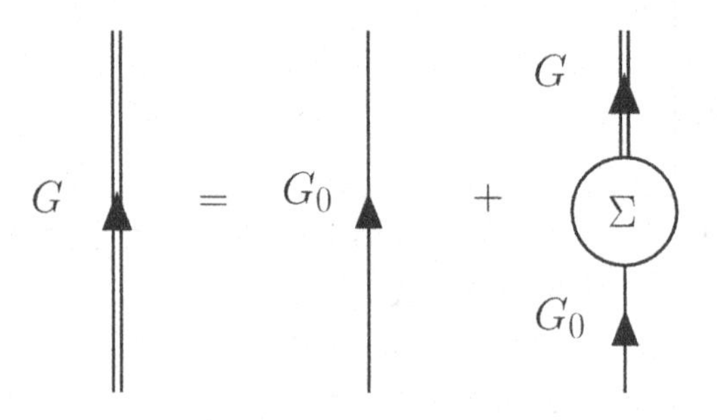

Figure 4.19 Diagrammatic representation of Dyson's equation for the nucleon propagator, Eq.(3.50).

Given an approximation scheme for Σ, Dyson's equation (4.112) is solved by iterations, and the Green's function is obtained in a fully self-consistent fashion.

To illustrate this procedure, let us neglect, for simplicity, three-nucleon forces. In this case, the operator T describing NN scattering in the nuclear medium is obtained summing up the infinite series of ladder diagrams represented in the top panel of Fig. 4.20. The T-matrix is in turn employed to derive the expression of the self-energy Σ, as illustrated in the bottom panel of the same figure, and the Green's function is obtained in a fully self-consistent fashion.

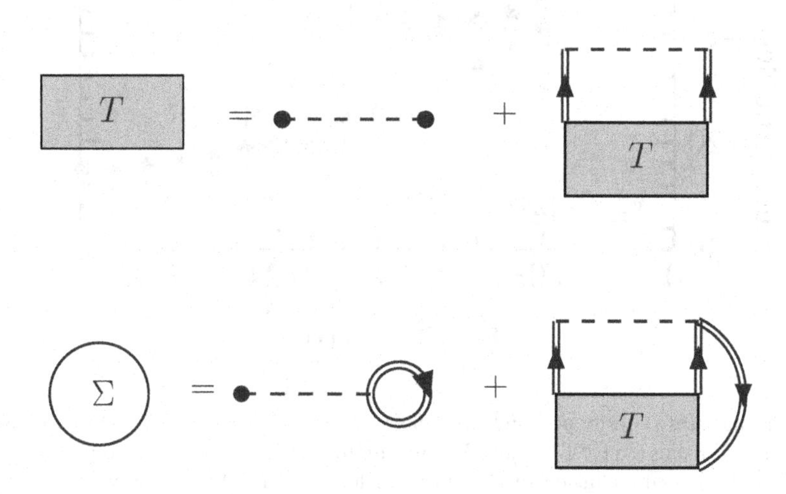

Figure 4.20 Diagrammatic representation of the T-matrix (top) and the self-energy (bottom) describing nucleons interacting through two-nucleon forces only. The dashed line and the box correspond to the NN potential and the T-matrix, respectively, while the double oriented line represents the dressed nucleon propagator.

Compared to the BHF approximation, in which only particle-particle ladder diagrams are taken into account, the SCGF approach allows to consistently include ladder diagrams involving hole-hole intermediate states. In addition, nucleon propagation is described by the dressed Green's function.

The SCGF approach has been generalised to include the contributions of three-nucleon forces, which are known to be essential in determining saturation of SNM and, more generally, the EOS of nuclear matter at supranuclear densities. The three-nucleon potential is reduced to a density-dependent two-nucleon potential—averaged over the position and quantum numbers of the third particle—and a new diagrammatic expansion, involving a large number of additional terms, is rewritten in terms of the resulting effective interaction.

In principle, the ground-state energy per particle can be obtained from the Green's function using Eq .(3.55). However, calculations are performed at non-vanishing temperature, to avoid the instability arising from nucleon pairing at low density.

Figure 4.21 shows the density dependence of the energy per nucleon of SNM at $T = 5$ MeV, computed using a χEFT Hamiltonian comprising a N^{3}LO NN potential and a density dependent potential derived from a N^{2}LO NNN potential [90]. Three-nucleon interactions appear to play a role even below equilibrium density, and become critical in shaping the EOS in the high-density region.

A very important feature of the SCGF formalism is the capability to describe nuclear

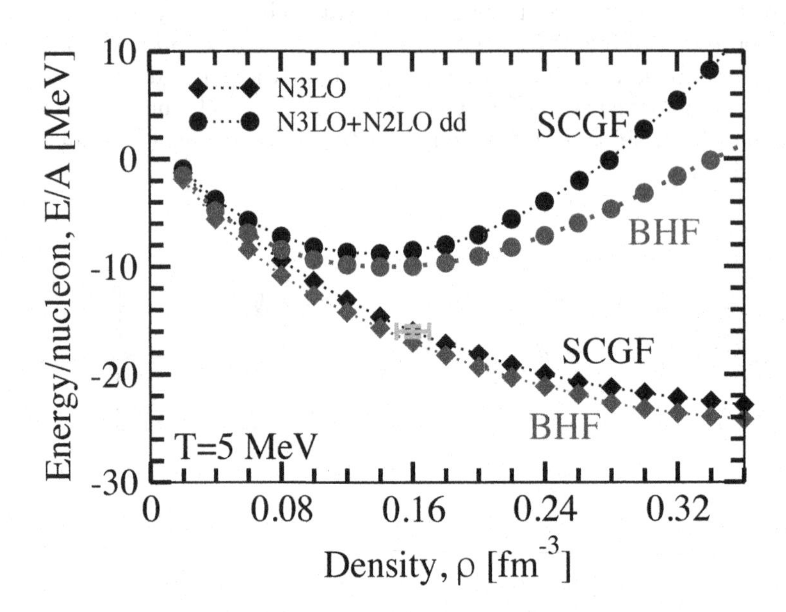

Figure 4.21 Ground-state energy per nucleon of SNM at temperature $T = 5$ MeV, computed within the SCGF approach using a χEFT nuclear Hamiltonian. Dots and diamonds represent the results obtained with and without inclusion of the three-nucleon potential. For comparison, the energies computed in the BHF approximation are also shown. Taken from [90].

matter properties other that the EOS, that can in principle be exploited to constrain the underlying dynamical model.

At energies $\omega < \mu$, with μ being the chemical potential, the spectral function $P(\mathbf{k}, \omega)$—trivially related to the Green's function through Eq. (3.54)—describes the probability of removing a nucleon from the nuclear ground state leaving the residual system with excitation energy ω. The spectral functions of a variety of nuclei have been obtained from high-resolution measurements of the cross sections of nucleon knock out $(e, e'p)$ processes [50]. The result of accurate nuclear matter calculations provide essential information needed for the interpretation of these experimental data [53].

Figure 4.22 shows the spectral function, defined in Eq.(3.54), obtained within the SCGF approach using the same χEFT Hamiltonian employed to determine the energies of Fig. 4.21. The peaks corresponding to quasiparticle states are clearly visible, and their width becomes vanishingly small at the Fermi surface—corresponding to $\omega = \mu$, with μ being the chemical potential—as predicted by Landau's theory. For $k < k_F$ the peaks are located in the region of $\omega - \mu < 0$, and correspond to hole states, whereas the peaks associated with particle states with $k > k_F$ occur at energies such that $\omega - \mu > 0$. In addition to the peaks, whose position and width can be explained by mean field dynamics, the spectral function exhibits a smooth background, extending to large values of $|\omega - \mu|$, arising from NN correlations. For comparison, consider that in the non interacting Fermi gas the spectral function reduces to a collection of δ-function peaks, located at $\omega = (k^2 - k_F^2)/2m$.

The effects of three-nucleon interactions are illustrated by the differences between the solid and dashed lines. They appear to be nearly negligible at the Fermi surface, independent

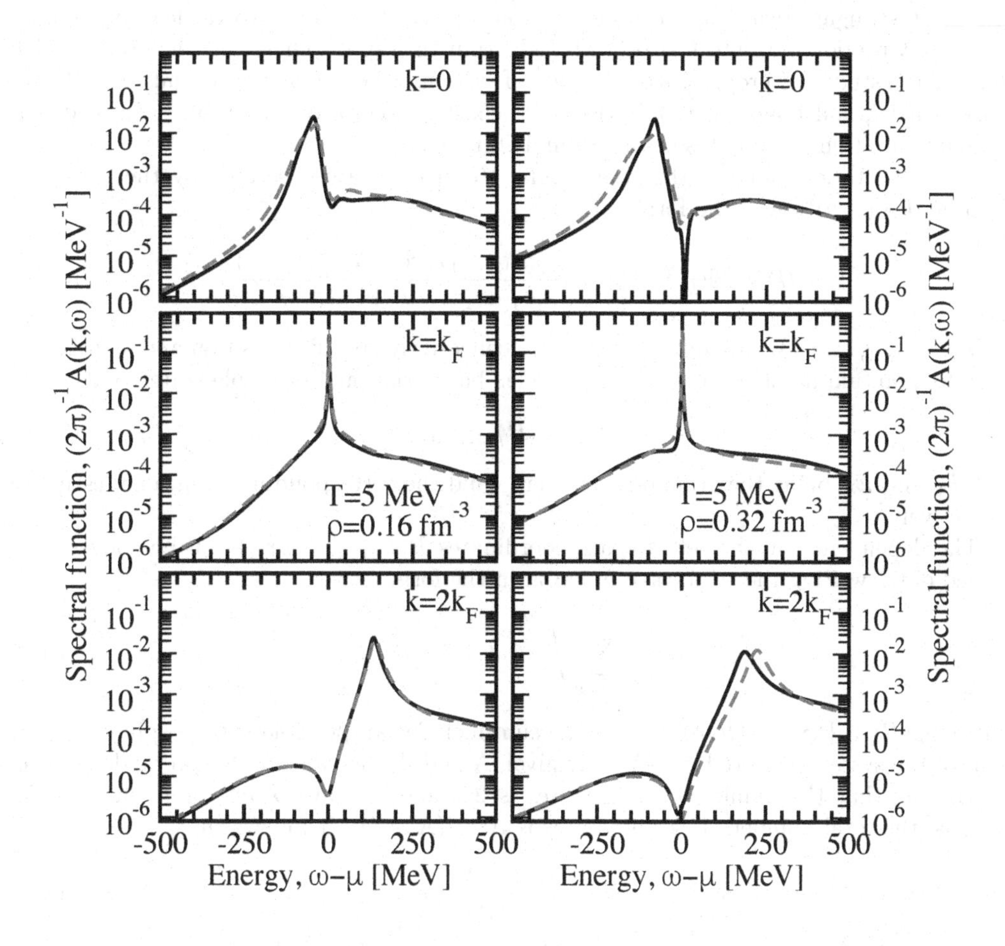

Figure 4.22 Spectral function of SNM at temperature $T = 5$ MeV, computed within the SCGF approach using a χEFT Hamiltonian comprising two- and three-nucleon potentials. The solid line show the results of the full calculation, while the dashed lines have been obtained taking into account two-nucleon forces only. Taken from [90].

of density. On the other, sizeable effects are visible at large energy and momentum for the higher density, $\varrho = 2\varrho_0$.

4.6 MONTE CARLO METHODS

As pointed out in Section 4.4, the evaluation of multidimensional integrals is a critical issue of the approach based on correlated wave functions, which is only marginally alleviated by the use of the cluster expansion formalism at finite order. Starting from the late 1970s, the stochastic approach known as Monte Carlo method [91, 92] has emerged as a powerful framework, capable to tackle this problem, as well as to overcome the inherent limitations of the variational approach, arising from the choice of the trial wave function.

4.6.1 Variational Monte Carlo

The most straightforward application of Monte Carlo techniques to nuclear many-body theory is Variational Monte Carlo (VMC). Within this approach—which has been widely employed to study the properties of a variety of systems other than nuclear matter, including atoms, nuclei, and liquid ^{3}He and ^{4}He—a stochastic integration algorithm is exploited to compute variational estimates of the ground-state energy.

The expectation value of the Hamiltonian H in the state described by the trial wave function Ψ_T is written in the form

$$\langle H \rangle = \langle \Psi_T | H | \Psi_T \rangle = \frac{\int dx_1 \ldots dx_{\rm A} \Psi_T^\dagger(X) H \Psi_T(X)}{\int dx_1 \ldots dx_{\rm A} |\Psi_T(X)|^2} , \tag{4.113}$$

with $X \equiv \{x_1, \ldots x_{\rm A}\}$, where $x_i \equiv \{\mathbf{r}_i, \sigma_i, \tau_i\}$ collectively specifies position and spin-isospin discrete quantum numbers of the i-th particle. The variational principle states that

$$E_V = \langle H \rangle \geq E_0 , \tag{4.114}$$

with E_0 and E_V being the ground-state energy and the corresponding variational estimate, respectively.

The dependence on the set of space coordinates $R \equiv \{\mathbf{r}_1, \ldots, \mathbf{r}_A\}$ can be conveniently singled out rewriting the trial wave function in the form

$$|\Psi_T\rangle = \sum_{n=1}^{M} \int dR \; \Phi_n(R) \; |R\rangle |S_n\rangle , \tag{4.115}$$

where $\Phi_n(R) = \{\langle S_n | \otimes \langle R |\} |\Psi_T\rangle$ and the sum includes all possible spin-isospin configurations of the system, denoted S_n. For any given A and Z, the value of M can be determined considering that the number of spin states is $2^{\rm A}$, and, because Z out of A nucleons are protons, the total number of isospin states is A!/Z!(A – Z)!. It follows that

$$M = 2^A \frac{A!}{Z!(A-Z)!} . \tag{4.116}$$

In the representation of Eq.(4.115), the Hamiltonian H is a $M \times M$ matrix, whose elements $H_{mn} = \langle S_m | H | S_n \rangle$ are functions of R. Substitution of Eq.(4.115) into Eq.(4.113) yields

$$\langle H \rangle = \frac{\int dR \; \sum_{m,n} \Phi_m^\star(R) H_{mn}(R) \Phi_n(R)}{\int dR \; \sum_n |\Phi_n(R)|^2} . \tag{4.117}$$

The expectation value $\langle H \rangle$ can be further rewritten in the form

$$\langle H \rangle = \int dR \; E(R) P(R) , \tag{4.118}$$

with the local energy $E(R)$ given by

$$E(R) = \frac{\sum_{m,n} \Phi_m^\star(R) H_{mn}(R) \Phi_n(R)}{\sum_n |\Phi_n(R)|^2} , \tag{4.119}$$

and

$$P(R) = \frac{\sum_n |\Phi_n(R)|^2}{\int dR \; \sum_n |\Phi_n(R)|^2} . \tag{4.120}$$

Let us now consider a set $\{R\} \equiv \{R_1, \dots, R_{N_c}\}$ of N_c configurations sampled according to the probability distribution $P(R)$, that is, such that the probability that a configuration R_i belong to the set $\{R\}$ is proportional to $P(R_i)$. It then follows that

$$\langle H \rangle = \lim_{N_c \to \infty} \frac{1}{N_c} \sum_{R_i \in \{R\}} E(R_i) , \tag{4.121}$$

and $\langle H \rangle$ can be estimated by truncating the infinite sum appearing in the right-hand side of the above equation at some large but finite value of N_c.

In principle, the VMC algorithm—in which the multidimensional integration is performed by sampling the 3A space coordinates of the nucleons—may be used to obtain accurate variational upper bounds to the ground-state energy of nuclear matter, without resorting to the cluster expansion. However, owing to the sum over spin and isospin degrees of freedom, the computational effort required to obtain $\langle H \rangle$ from Eq.(4.121) is a rapidly growing function of A and Z, see Eq.(4.116). This problem severely hampers the application of VMC to nuclear matter, modelled as a collection of nucleons confined to a box with periodic boundary conditions. Numerical studies are so far limited to PNM, and include 14 particles [93].

The obvious method to avoid performing the full spin-isospin sum is sampling the spin-isospin configurations of the system, S_n. Within this scheme, Eqs.(4.118)-(4.120) are replaced by

$$\langle H \rangle = \int dR \sum_n E_n(R) P_n(R) , \tag{4.122}$$

where

$$E_n(R) = \frac{\sum_m \Phi^\star_m(R) H_{mn}(R) \Phi_n(R)}{\sum_m |\Phi_m(R)|^2} , \tag{4.123}$$

and

$$P_n(R) = \frac{|\Phi_n(R)|^2}{\int dR \sum_m |\Phi_m(R)|^2} . \tag{4.124}$$

Conceptually, this route is pursued by the Auxiliary Field Diffusion Monte Carlo (AFDMC) approach, to be discussed in Section 4.6.2.

A further, and potentially serious, limitation of the VMC approach is related to the choice of the trial wave function, that should ideally embody all effects of nuclear dynamics, at both short and long range.

4.6.2 Auxiliary field diffusion Monte Carlo

The Auxiliary Field Diffusion Monte Carlo (AFDMC) method is based on the Diffusion Monte Carlo (DMC) method, which allows to obtain the ground state of the Hamiltonian from the solution of the imaginary time Schrödinger equation

$$-\frac{\partial \Psi(X,\tau)}{\partial \tau} = H \Psi(X,\tau) , \tag{4.125}$$

where $\tau = it$ and X specifies the particle positions, spins and isospins. The solution of the above equation can be written in the form

$$\Psi(X,\tau) = \int dX' G(X,X',\tau) \Psi(X',0) , \tag{4.126}$$

with the kernel $G(X, X', \tau)$ being given by

$$G(X, X', \tau) = \langle X | e^{-H\tau} | X' \rangle \ . \tag{4.127}$$

In principle, solving Eq.(4.125) allows to overcome the deficiencies of the variational wave function, by enhancing its projection onto the true ground-state of the system. To see this, let us consider the expansion of the trial wave function in the basis of eigenstates of the Hamiltonian

$$|\Psi_T\rangle = \sum_n a_n |\Psi_n\rangle \ , \tag{4.128}$$

and set $|\Psi(0)\rangle = |\Psi_T\rangle$. Assuming that $|\Psi_T\rangle$ is not orthogonal to the ground state $|\Psi_0\rangle$[4], it follows that

$$\begin{aligned} \lim_{\tau\to\infty} |\Psi(\tau)\rangle &= \lim_{\tau\to\infty} e^{-(H-E_T)\tau} \sum_n a_n |\Psi_n\rangle \\ &= \lim_{\tau\to\infty} e^{-(E_0-E_T)\tau} \sum_n e^{-(E_n-E_0)\tau} a_n |\Psi_n\rangle \ , \end{aligned} \tag{4.129}$$

with $E_n - E_0 > 0$, implying

$$\lim_{\tau\to\infty} |\Psi(\tau)\rangle = \lim_{\tau\to\infty} e^{-(E_0-E_T)\tau} a_0 |\Psi_0\rangle \ . \tag{4.130}$$

The above relation, in which the offset parameter E_T can be chosen to be the ground-state energy obtained from a variational calculation, shows that in the large τ limit the solution of Eq.(4.125) reduces to the ground state wave function.

Within the DMC approach, the function $\Psi(X, \tau)$ is obtained by dividing the imaginary time τ into N short intervals $\Delta\tau$, and iterating the short-time evolution of a population of configurations of the system, dubbed walkers, dictated by the equation

$$\Psi(X, \tau + \Delta\tau) = \int dX' \langle X | e^{-(T+V)\Delta\tau} | X' \rangle \Psi(X', \tau) \ , \tag{4.131}$$

where T and V denote the kinetic and potential energy terms of the Hamiltonian. For small $\Delta\tau$ one can apply Trotter's formula to write

$$e^{-(T+V)\Delta\tau} = e^{-T\Delta\tau} e^{-V\Delta\tau} + \mathcal{O}(\Delta\tau^2) \ , \tag{4.132}$$

and recast the kernel of Eq.(4.131) in the factorised form

$$G(X, X', \Delta\tau) = G_0(R, R', \Delta\tau) W(X, X', \Delta\tau) \ , \tag{4.133}$$

with

$$G_0(R, R', \Delta\tau) = \left(\frac{m}{2\pi\Delta\tau} \right)^{3A/2} e^{-m(R-R')^2/2\Delta\tau} \ , \tag{4.134}$$

$$W(X, X', \Delta\tau) = e^{-[V(X)-E_T]\Delta\tau} \ . \tag{4.135}$$

The 3A–dimensional integration involved in Eq.(4.131) is performed using a Monte Carlo algorithm. The particle coordinates at $\tau = 0$ are sampled from the probability distribution

[4]This assumption amounts to requiring that $a_0 \neq 0$.

associated with the variational wave function, while the evolution is driven by the Gaussian diffusion term G_0 and the weight factor W.

Because the integration in Eq.(4.131) also involves the spin–isospin degrees of freedom, a DMC calculation with a realistic nuclear Hamiltonian—including spin–isospin dependent potentials—requires a sum over all possible spin-isospin states of the system entering the definition of the trial wave function, see Eqs. (4.115) and (4.116). This procedure is followed in the Green's Function Monte Carlo (GFMC) method, first proposed by M. Kalos in 1962 [94] and successfully applied in condensed matter physics and chemistry. However, its extension to the treatment of nuclear systems is very demanding in terms of computational resources. Owing to this limitation, GFMC results are only available for nuclei with A $\leq$ 12 [22] and PNM with 14 neutrons in a periodic box [93].

Within AFDMC the above problem—arising from the occurrence of terms quadratic in the nucleon spin and isospin in the potential $V(X)$—is circumvented performing a Hubbard–Stratonovich transformation, that reduces the quadratic dependences to a linear dependence.

For a generic spin-dependent operator O, the tranformation can be written in the form

$$e^{-\frac{1}{2}\lambda O^2 \tau} = \frac{1}{\sqrt{2\pi}} \int dx \; e^{-\frac{1}{2}x^2 + x\sqrt{-\lambda\tau}O} \; , \tag{4.136}$$

with λ a parameter.

The linearisation of the operator dependence of the potential leads to a great deal of simplification, because it allows the use of a basis of single-nucleon spinors. Within this scheme, a spin-isospin configuration of the system is specified by the outer product

$$[c_1^1|p\uparrow\rangle + c_1^2|p\downarrow\rangle + c_1^3|n\uparrow\rangle + c_1^4|n\downarrow\rangle] \otimes [c_2^1|p\uparrow\rangle + c_2^2|p\downarrow\rangle + c_2^3|n\uparrow\rangle + c_2^4|n\downarrow\rangle] \otimes \ldots \; ,$$

where the coefficients c_i^α denote the amplitudes to find the i–th nucleon in the spin–isospin state α. In the new basis, the number of configurations contributing to the spin-isospin sum of Eq.(4.131), that is, the dimension of the vector representing the system in spin-isospin space, is 4A, to be compared to M of Eq.(4.116). Therefore, the computational effort required by AFDMC calculations is significantly reduced, with respect to GFMC, already in the case of a light nucleus such as ^{4}He, in which case 4A = 16 and $M = 96$, see Eq.(4.116).

Within AFDMC, the Green's function describing the short-time evolution in spin-isospin space driven by a NN potential including the six operators of Eq.(2.13) is written in the form

$$\begin{aligned} G(R, R', \Delta\tau) &= G_0(R, R', \Delta\tau) e^{-(v_{SI}(R) - E_T)\Delta\tau} \\ &\times \prod_{n=1}^{15A} \int \frac{dx_n}{\sqrt{2\pi}} e^{-\frac{1}{2}x_n^2 + x\sqrt{-\lambda_n\tau}O_n} \; . \end{aligned} \tag{4.137}$$

where v_{SI} denotes the spin-isospin independent part of the potential, and the sum includes 3A + 3A contributions arising from the spin-spin and isospin–isospin components of the potential, and 9A contributions arising from the spin–isospin component. The integration in the second line of the above equation is performed by sampling the new variables x_n, called auxiliary fields.

It has been suggested that the auxiliary fields may also have a physical interpretation as meson fields responsible of NN Interactions. This interpretation emerges in the case of the OPE interaction, because the operators appearing in the Green's function of Eq.(4.137) have exactly the same structure as the interaction between a pion field and a nucleon [95].

Early applications of the AFDMC to nuclear matter have been performed using a somewhat simplified model of the nuclear Hamiltonian, in which the NNN potential was replaced by a density-dependent NN potential, designed to account for both three- and many-nucleon interactions [96]. More recently The AFDMC method has been employed to carry out accurate calculations of the properties of both SNM and PNM. However, studies of SNM require additional computational issues, and the existing results have been obtained neglecting the effects of three-nucleon interactions.

Figure 4.23 shows the ground-state energy per nucleon of SNM computed using the AFDMC method and using two different NN potentials. Simulations have been carried out including up to 132 nucleons in a periodic box'[97]. The lower curve marked with squares, showing a monotonically decreasing behaviour as a function of density, corresponds to the AV6P potential, see Chapter 2, while the results obtained from the v_7' (AV7P) model, in which the isospin-independent part of the spin-orbit interaction is also included, are represented by the curve marked by dots. It is apparent that the AV7P interaction is more repulsive at densities around and above nuclear matter equilibrium density, and predicts saturation, although the corresponding energy and density are both quite far from the empirical values.

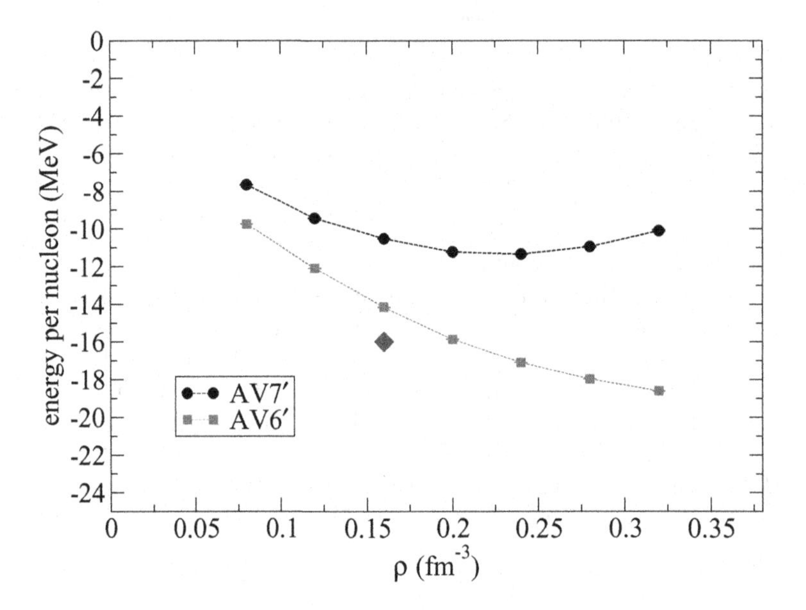

Figure 4.23 Ground-state energy per nucleon of SNM as a function of the density. The upper and lower curves represent AFDMC results obtained using the AV6P and AV7P NN potentials, respectively. Taken from [97].

In Fig. 4.24, the results of a calculations of the energy per nucleon of PNM are displayed as a function of matter density. Simulations have been performed using 66 neutrons in a periodic box and two different models of the nuclear Hamiltonian, in which interactions are described by the AV8P NN potential, with and without inclusion of the UIX NNN potential. The repulsive contribution arising from three-nucleon interactions is already visible around the empirical saturation density, and becomes very large in the high density regime of $\varrho \gtrsim 2\varrho_0$.

The AFDMC method has been also employed to PNM using nuclear hamiltonian derived

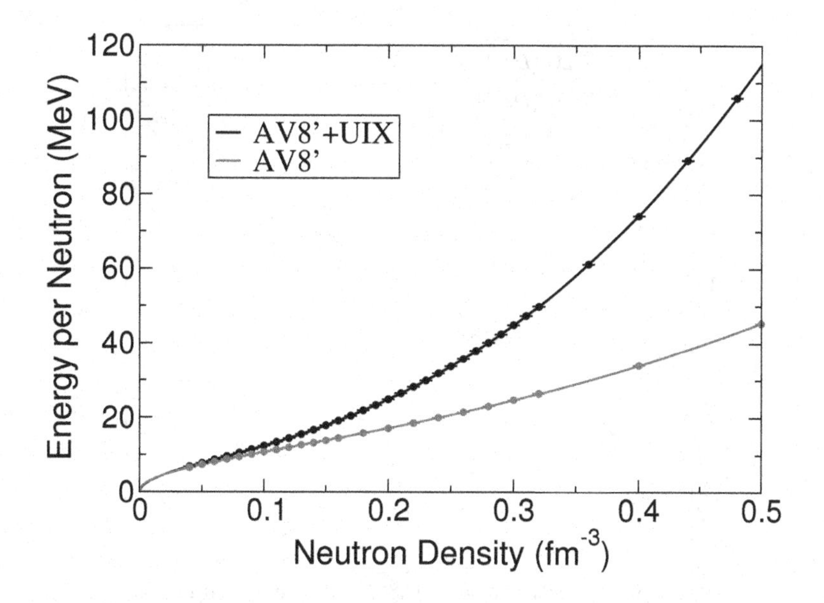

Figure 4.24 Ground-state energy per nucleon of PNM as a function of the density. The upper and lower curves represent AFDMC results obtained using the AV8P + UIX and AV8P nuclear Hamiltonians, respectively. Taken from [98].

from χEFT discussed in Chapter 2 [36, 99]. Figure 4.25 shows the results of calculations carried out using the local N^2LO NN potential of Gezerlis *et al.* corresponding to the coordinate-space cut off $R_0 = 1$ fm, see Section 2.5 [31, 32], supplemented with three different implementations of the N^2LO NNN potential, featuring different contact terms. The shaded region provides an indication of the combined uncertainties associated with the choice of NNN potential and the estimated size of higher order contributions. It clearly appears that the application of the dynamical model based on χEFT to nuclear matter is plagued by large uncertainties, even at density below the equilibrium density of SNM.

4.7 RELATIVITY

The problem of taking into account relativistic corrections to the description of nuclear matter has been attacked using different strategies. The proposed approaches involve modifications of the formalism of nuclear many-body theory, of the underlying models of nuclear interactions, or both.

4.7.1 Boost corrections to the nucleon-nucleon potential

Being designed to account for the measured NN scattering phase shifts, the phenomenological NN potentials discussed in Chapter 2 describe the interaction in the centre of mass frame, defined by the condition

$$\mathbf{P} = \mathbf{p}_i + \mathbf{p}_j = 0 \ , \tag{4.138}$$

where $\mathbf{P}$ is the total momentum of the participating nucleons, carrying momenta $\mathbf{p}_i$ and $\mathbf{p}_j$. In nuclear matter, however, NN collisions involve nucleon pairs having $\mathbf{P} \neq 0$, and are

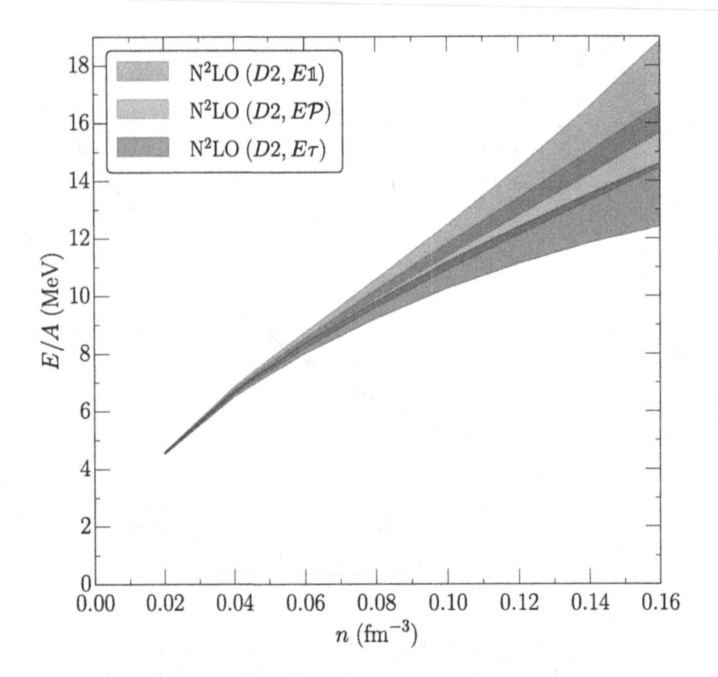

Figure 4.25 Ground-state energy per nucleon of PNM as a function of the density. The upper and lower curves represent AFDMC results obtained using a local N^2LO NN potential and three different implementations of the NNN potential. Taken from [36].

driven by a potential that can be written in the form

$$v_{ij}(\mathbf{P}) = v_{ij} + \delta v_{ij}(\mathbf{P}) \ . \tag{4.139}$$

In the above equation, $v_{ij} = v_{ij}(0)$ and $\delta v_{ij}(\mathbf{P})$, referred to as boost correction, accounts for the effects of centre of mass motion.

Within the scheme of Eq.(4.139), v_{ij} is determined by nuclear dynamics only, while $\delta v_{ij}(\mathbf{P})$, obviously satisfying the condition $\delta v_{ij}(0) = 0$, is derived from relativistic covariance.

The explicit expression of δv_{ij} including terms up to oder $\mathbf{P}^2/m^2$ was obtained by R. Krajcik and L. Foldy in the 1970s [100]. It can be written in the form [101]

$$\begin{aligned} \delta v_{ij}(\mathbf{P}) = & -\frac{\mathbf{P}^2}{8m^2} v_{ij} + \frac{1}{8m^2}[(\mathbf{P}\cdot\mathbf{r})(\mathbf{P}\cdot\boldsymbol{\nabla}), v_{ij}] \\ & + \frac{1}{8m^2}[(\boldsymbol{\sigma}_i - \boldsymbol{\sigma}_j)\times\mathbf{P}\cdot\boldsymbol{\nabla}, v_{ij}], \end{aligned} \tag{4.140}$$

where the operator $\boldsymbol{\sigma}_i$ describes the spin of nucleon i, and $\boldsymbol{\nabla}$ acts on the relative coordinate $\mathbf{r} = \mathbf{r}_i - \mathbf{r}_j$.

The results of VMC studies of the few-nucleon systems have shown that inclusion of the boost interaction in the nuclear Hamiltonian results in the appearance of a sizeable repulsive contribution to the binding energies of ^{3}He and ^{4}He [102]. As a consequence the repulsion arising from the three-body potential V^R_{ijk} of Eq.(2.15), needed to reproduce the experimental values of the energies, turns out to be reduced.

A simplified expression of the relativistic boost correction δv_{ij}, obtained neglecting the last term in the right-hand side of Eq.(4.140), as well as the non static components of v_{ij}, has been employed in advanced variational calculations of the EOS of both PNM and SNM. The results of these studies will be discussed in Chapter 5.

4.7.2 Dirac-Brueckner formalism

A relativistic treatment suitable to perform calculations of proton-nucleus scattering observables was first proposed at the end of the 1970s [103]. Within this approach, the projectile particle is described by the Dirac equation

$$[\not{p} - (m + U_S) - \gamma^0 (U_V + V_C)]\psi_{\mathbf{p}}(r) , \tag{4.141}$$

where $p \equiv (E, \mathbf{p})$, with E and $\mathbf{p}$ being the proton energy and momentum. In the above equation, U_S and U_V denote a Lorentz scalar potential and the time component of a four-vector potential, whereas V_C is the potential describing the Coulomb interaction with the electrostatic field of the target nucleus. The Lorentz structure of the potential is inspired by the arguments underlying the derivation of the OBE potentials, see Section 2.4.

The success of the approach based on Eq. (4.141), referred to as Dirac's phenomenology, motivated a great deal of effort aimed at extending the relativistic formalism based on Dirac's equation to study the properties of nuclear matter [104]. It has to be pointed out, however, that, unlike the analyses of proton-nucleus collisions, which are based on phenomenological potentials to be determined from a fit to experimental data, nuclear matter calculations do not involve any adjustable parameters.

The generalisation of BBG theory to take into account relativistic effects, referred to as Dirac-Brueckner theory, is based on the Dirac equation describing a single nucleon carrying four-momentum p [105], to be compared to Eq.(4.141)

$$(\not{p} - m - U)\widetilde{u}_s(\mathbf{p}) , \tag{4.142}$$

where

$$U = U_S + \gamma^0 U_V , \tag{4.143}$$

and

$$\widetilde{u}_s(\mathbf{p}) = \sqrt{\frac{E^\star_{\mathbf{p}} + m^\star}{2m^\star}} \begin{pmatrix} \chi_s \\ \dfrac{(\boldsymbol{\sigma} \cdot \mathbf{p})}{E^\star_{\mathbf{p}} + m^\star}\chi_s \end{pmatrix} . \tag{4.144}$$

Here, $m^\star = m + U_s$ and $E^\star_{\mathbf{p}} = \sqrt{m^{\star 2} + \mathbf{p}^2}$, while χ_s denotes a two-component Pauli spinor.

The extension of BBG theory requires an operator describing NN scattering in the nuclear medium, analog to the G-matrix of non relativistic theory. The relativistic G-matrix is obtained from a three-dimensional reduction of the Bethe-Salpeter equation, referred to as Thompson equation, that can be written in the form [105]

$$\begin{aligned} \langle \mathbf{k}|G(W)|\mathbf{k}\rangle &= \langle \mathbf{k}|\tilde{v}|\mathbf{k}\rangle \\ &+ \int \frac{d^3k'}{(2\pi)^3} \left(\frac{m^\star}{E^\star_{\mathbf{k}'+\mathbf{K}/2}}\right)^2 \langle \mathbf{k}|\tilde{v}|\mathbf{k}'\rangle \frac{f(\mathbf{k}', \mathbf{K})}{W - e^\star_{\mathbf{k}'+\mathbf{K}/2} - e^\star_{\mathbf{k}'-\mathbf{K}/2}} \langle \mathbf{k}'|G(W)|\mathbf{k}\rangle , \end{aligned} \tag{4.145}$$

with

$$e^\star_{\mathbf{k}} = E^\star_{\mathbf{k}} + U_V , \tag{4.146}$$

and

$$W = e^\star_{\mathbf{k}+\mathbf{K}/2} + e^\star_{\mathbf{k}-\mathbf{K}/2} , \tag{4.147}$$

to be compared to Eq.(4.27). In the above equation, $\mathbf{K}$ is the centre of mass momentum, $\mathbf{k}$ and $\mathbf{k}'$ are the initial and intermediate relative momenta, respectively, and the effect of Pauli's principle is taken into account by the function $f(\mathbf{k}', \mathbf{K})$, defined by Eq.(4.29).

The calculation of the matrix element $\langle \mathbf{k}|\tilde{v}|\mathbf{k}\rangle$ is carried out using a OBE potential and the spinors of Eq.(4.144). As a consequence, it depends strongly on density through the effective mass $m^\star$. This is the most important difference between Eq.(4.145) and the corresponding non relativistic expression, Eq.(4.27).

The ground-state energy of nuclear matter is obtained from

$$\begin{aligned} E_0 = & \sum_{\mathbf{k}_i \in \{F\}} \frac{m^\star}{E^\star_{\mathbf{k}_i}} \langle \mathbf{k}_i | \gamma \cdot \mathbf{k}_i + m | \mathbf{k}_i \rangle \\ & + \sum_{\mathbf{k}_i, \mathbf{k}_j \in \{F\}} \frac{m^{\star 2}}{E^\star_{\mathbf{k}_i} E^\star_{\mathbf{k}_j}} \langle \mathbf{k}_i \mathbf{k}_j | G(e^\star_{\mathbf{k}_i} + e^\star_{\mathbf{k}_j}) | \mathbf{k}_i \mathbf{k}_j \rangle_A - m , \end{aligned} \tag{4.148}$$

where the index A indicate that the state $|\mathbf{k}_i \mathbf{k}_j\rangle$ is properly antisymmetrised.

The single-nucleon potential is obtained self-consistently from

$$U_{\mathbf{k}_i} = \frac{m^\star}{E^\star_{\mathbf{k}_i}} \langle \mathbf{k}_i | U_S + \gamma^0 U_V | \mathbf{k}_i \rangle = \text{Re} \sum_{\mathbf{k}_j \in \{F\}} \langle \mathbf{k}_i \mathbf{k}_j | G(e^\star_{\mathbf{k}_i} + e^\star_{\mathbf{k}_j}) | \mathbf{k}_i \mathbf{k}_j \rangle_A , \tag{4.149}$$

for all values of $|\mathbf{k}_i|$. The scheme described above is referred to as Dirac-Brueckner-Hartree-Fock, or DBHF, approximation.

Figure 4.26 shows the energy per nucleon of SNM computed within the DBHF approximation using the OBE Bonn potentials labelled model A, B, and C [105]. A comparison with the corresponding results obtained from the non relativistic theory shows that, independent of the potential model, the inclusion of relativistic effects results in a sizeable repulsive correction to the energy. As a result, the saturation points move to lower Fermi momenta and higher energies, and the agreement with the empirical data is significantly improved.

The observation that the repulsive relativistic effect is similar to the effect of three-nucleon forces, see Fig. 4.6, suggests that the DBHF approximation allows to take into account mechanisms leading to the appearance of three-nucleon interaction.

Consider the Z-graph of Fig. 4.27, in which oriented lines pointing downward or upward correspond to nucleons and antinucleons, and dashed lines represent the NN potential. The intermediate state features a nucleon-antinucleon pair—arising from the expansion of the spinor of Eq.(4.144) in positive and negative energy solutions of the Dirac equation in free space—that gives rise to the appearance of a three-nucleon force [107].

4.7.3 Relativistic mean-field approximation

The theoretical approaches described in the previous sections are based on the assumption that the degrees of freedom associated with the carriers of the interaction between nucleons can be eliminated in favour of potentials. This scheme has proved to be very useful to describe the properties of atomic nuclei, and is therefore expected to be applicable to nuclear matter up to densities $\varrho \sim \varrho_0$. As the density increases, however, the relativistic propagation of the nucleons, as well as the retarded propagation of the virtual meson fields giving rise to nuclear forces, are expected to become more and more important.

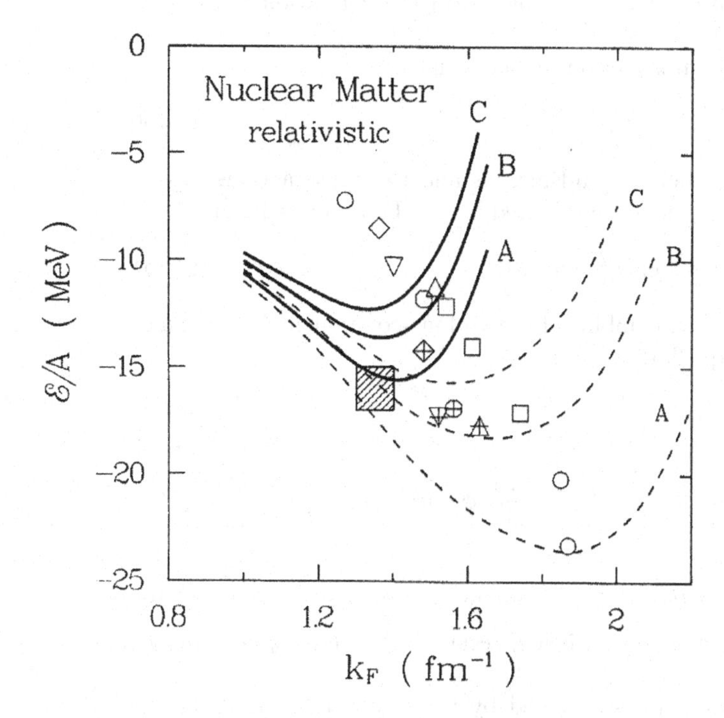

ıre 4.26 Ground-state energy per nucleon of SNM as a function of the Fermi momentum. The d and dashed lines represent DBHF and BHF results, obtained using the Bonn OBE potentials B and C. For comparison, the saturation points predicted by other BHF calculations are also layed by the symbols. The shaded square highlights the region compatible with the empirical rmation on saturation of SNM.Taken from [106].

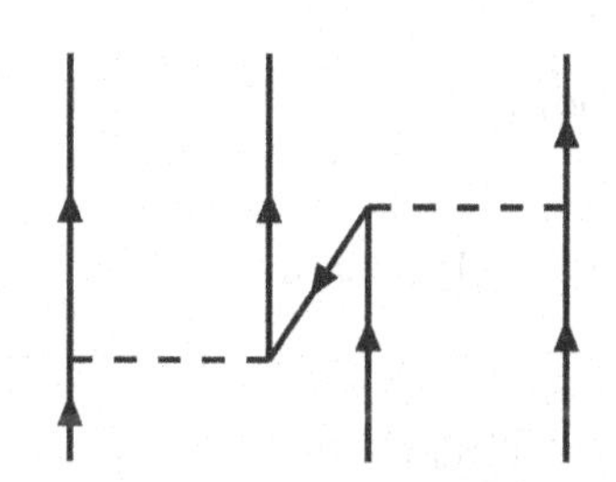

ıre 4.27 Diagrammatic representation of a three-nucleon force arising from the excitation of a ual nuclon-antinucleon pair. Oriented lines pointing downward or upward correspond to nucleons antinucleons, respectively, while dashed lines depict the NN potential.

Within the model first proposed by J. Walecka in the 1970s, nuclear matter consists of leons, described by Dirac spinors, interacting through exchange of a scalar and a vector ıeson, called σ and ω, reminiscent of the mesons employed in the derivation of the OBE ·otentials, see Section 2.4 [108].

The Lagrangian density of the system is written in the form

$$\mathcal{L} = \mathcal{L}_N + \mathcal{L}_B + \mathcal{L}_{int} \ , \tag{4.150}$$

/here $\mathcal{L}_N$, $\mathcal{L}_B$ and $\mathcal{L}_{int}$ describe free nucleons and mesons and their interactions, respec-ively. The dynamics of the free nucleon field are dictated by the Dirac Lagrangian (2.4)

$$\mathcal{L}_N(x) = \bar{\psi}(x) \left(i\partial\!\!\!/ - m\right) \psi(x) \ , \tag{4.151}$$

/here the nucleon field, denoted by $\psi(x)$, combines the two four-component Dirac spinors escribing proton and neutron, see Eq. (2.5).

The meson Lagrangian reads

$$\begin{aligned} \mathcal{L}_B(x) &= \mathcal{L}_\omega(x) + \mathcal{L}_\sigma(x) \\ &= -\frac{1}{4}F^{\mu\nu}(x)F_{\mu\nu}(x) + \frac{1}{2}m_\omega^2 V_\mu(x)V^\mu(x) + \frac{1}{2}\partial_\mu\phi(x)\partial^\mu\phi(x) - \frac{1}{2}m_\sigma^2\phi(x)^2 \ , \end{aligned} \tag{4.152}$$

/here

$$F_{\mu\nu}(x) = \partial_\nu V_\mu(x) - \partial_\mu V_\nu(x) \ , \tag{4.153}$$

$V_\mu(x)$ and $\sigma(x)$ are the vector and scalar meson fields, respectively, and m_ω and m_σ the orresponding masses.

The form of the interaction Lagrangian is specified by the requirement that, besides eing a Lorentz scalar, $\mathcal{L}_{int}(x)$ give rise to a Yukawa meson-exchange potential in the static imit. The resulting expression is

$$\mathcal{L}_{int}(x) = g_\sigma\phi(x)\bar{\psi}(x)\psi(x) - g_\omega V_\mu(x)\bar{\psi}(x)\gamma^\mu\psi(x) \ , \tag{4.154}$$

/here g_σ and g_ω are coupling constants and the choice of signs reflect the observation that he NN interaction is known to comprise both attractive and repulsive contributions.

The equations of motion are obtained from minimisation of the action associated with he Lagrangian of Eq. (4.150). The meson fields satisfy

$$(\Box + m_\sigma^2)\phi(x) = g_\sigma \ \bar{\psi}(x)\psi(x) \tag{4.155}$$

ınd

$$(\Box + m_\omega^2)V_\mu(x) - \partial_\mu(\partial^\nu V_\nu) = g_\omega \ \bar{\psi}(x)\gamma_\mu\psi(x) \ , \tag{4.156}$$

/hile for the nucleon field one finds

$$\left[(\partial\!\!\!/ - g_\omega\gamma_\mu V^\mu(x)) - (m - g_\sigma\phi(x))\right]\psi(x) = 0 \ . \tag{4.157}$$

The above coupled equations are fully relativistic and Lorentz covariant. However, their olution involves prohibitive difficulties.

In most applications of the model, Eqs. (4.155)-(4.157) are drastically simplified by ntroducing the Mean Field Approximation (MFA), which amounts to replacing

$$\phi(x) \to \langle\phi(x)\rangle \ , \quad V_\mu(x) \to \langle V_\mu(x)\rangle \ , \tag{4.158}$$

/here $\langle\phi(x)\rangle$ and $\langle V_\mu(x)\rangle$ are the ground-state expectation values of the fields.

In nuclear matter, the baryon and scalar densities, $\varrho_B = \psi^\dagger\psi$ and $\varrho_S = \bar{\psi}\psi$, as well ıs the current $j_\mu = \bar{\psi}\gamma_\mu\psi$, are independent of x, and so are the expectation values of the son fields. In addition, rotation invariance implies that $\langle\bar{\psi}\gamma_i\psi\rangle = 0$, with $i = 1, 2, 3$. It ows that

$$m_\sigma^2 \langle\phi\rangle = g_\sigma\langle\bar{\psi}\psi\rangle \ , \tag{4.159}$$

$$m_\omega^2 \langle V_0\rangle = g_\omega\langle\psi^\dagger\psi\rangle \ , \tag{4.160}$$

$$m_\omega^2 \langle V_i\rangle = g_\omega\langle\bar{\psi}\gamma_i\psi\rangle = 0 \ , \tag{4.161}$$

the nucleon equation of motion reduces to

$$\left[(i\partial\!\!\!/ - g_\omega\gamma_\mu\langle V^\mu\rangle) - (m - g_\sigma\langle\phi\rangle)\right]\psi(x) = 0 \ . \tag{4.162}$$

Using the above results, the Lagrangian defined by Eqs.(4.150)-(4.152) and (4.154) can rewritten in the form

$$\mathcal{L}_{MFA} = \bar{\psi}\left[i\partial\!\!\!/ - g_\omega\gamma^0\langle V_0\rangle - (m - g_\sigma\langle\phi\rangle)\right]\psi - \frac{1}{2}m_\sigma^2\langle\phi\rangle^2 + \frac{1}{2}m_\omega^2\langle V_0\rangle^2 \ , \tag{4.163}$$

the associated energy-momentum tensor turns out to be

$$T_{MFA}^{\mu\nu} = i\bar{\psi}\gamma^\mu\partial^\nu\psi - g^{\mu\nu}\left[-\frac{1}{2}m_\sigma^2\langle\phi\rangle^2 - \frac{1}{2}m_\omega^2\langle V_0\rangle^2\right] \ . \tag{4.164}$$

en the energy-momentum tensor, the formalism of quantum field theory provides a sistent procedure to obtain the zero-temperature EOS, that, is the relation $P = P(\epsilon)$, re P and ϵ are the matter pressure and mass-energy density.

In nuclear matter, the ground-state expectation value of $T^{\mu\nu}$ is simply related to P and rough

$$\langle T^{\mu\nu}\rangle = u^\mu u^\nu\,(\epsilon + P) - g^{\mu\nu}P \ , \tag{4.165}$$

re u^μ denotes the four-velocity of the system, satisfying $u_\mu u^\mu = 1$. It follows that in the rence frame in which matter is at rest $\langle T^{\mu\nu}\rangle$ is diagonal, with

$$\langle T^{00}\rangle = \epsilon \quad , \quad \frac{1}{3}\langle T^{ii}\rangle = P \ . \tag{4.166}$$

llecting the above results, and using

$$\langle\bar{\psi}\gamma_0 k_0\psi\rangle = \frac{g_\omega^2}{m_\omega^2}\varrho_B^2 + \frac{\nu}{(2\pi)^3}\int d^3k\,\sqrt{\mathbf{k}^2 + (m - g_\sigma\langle\phi\rangle)^2}\;\theta(k_F - |\mathbf{k}|) \ , \tag{4.167}$$

$$\langle\bar{\psi}\gamma_i k_i\psi\rangle = \frac{\nu}{(2\pi)^3}\int d^3k\frac{|\mathbf{k}|^2}{\sqrt{|\mathbf{k}|^2 + (m - g_\sigma\langle\phi\rangle)^2}}\;\theta(k_F - |\mathbf{k}|) \ , \tag{4.168}$$

finally arrives at

$$\epsilon = \frac{1}{2}\frac{m_\sigma^2}{g_\sigma^2}(m - m^*)^2 + \frac{1}{2}\frac{g_\omega^2}{m_\omega^2}\varrho_B^2 + \frac{\nu}{2\pi^2}\int_0^{k_F}|\mathbf{k}|^2 d|\mathbf{k}|\;\sqrt{|\mathbf{k}|^2 + m^{*2}} \ , \tag{4.169}$$

$$P = -\frac{1}{2}\frac{m_\sigma^2}{g_\sigma^2}(m - m^*)^2 + \frac{1}{2}\frac{g_\omega^2}{m_\omega^2}\varrho_B^2 + \frac{1}{3}\,\frac{\nu}{2\pi^2}\int_0^{k_F}d|\mathbf{k}|\;\frac{|\mathbf{k}|^4}{\sqrt{|\mathbf{k}|^2 + m^{*2}}} \ . \tag{4.170}$$

e first two contributions to the right-hand side of the above equations arise from the ıass terms associated with the vector and scalar fields, while the remaining term gives the nergy density and pressure of a relativistic Fermi gas of nucleons of mass m^* given by

$$\begin{aligned} m^* &= m - \frac{g_\sigma^2}{m_\sigma^2}\frac{\nu}{2\pi^2}\int_0^{k_F}|\mathbf{k}|^2 d|\mathbf{k}| \frac{m^*}{\sqrt{|\mathbf{k}|^2+m^{*2}}} \\ &= m - \frac{g_\sigma^2}{m_\sigma^2}\frac{m^*}{\pi^2}\left[k_F e_F^* - m^{*2}\ln\left(\frac{k_F+e_F^*}{m^*}\right)\right] , \end{aligned} \tag{4.171}$$

ith $e_F^* = \sqrt{k_F^2+m^{*2}}$. Equations (4.169)-(4.171) yield energy density and pressure of uclear matter as a function of the baryon density $\varrho_B = \nu k_F^3/6\pi^2$. The values of the unnown quantities (m_σ^2/g_σ^2) and (m_ω^2/g_ω^2) are determined by a fit to the empirical saturation roperties of nuclear matter, that is, requiring

$$\frac{E_0}{A} = \frac{\epsilon(\varrho_0)}{\varrho_0} - m = -16 \text{ MeV} , \tag{4.172}$$

ith $\varrho_0 = .16$ fm^{-3}.

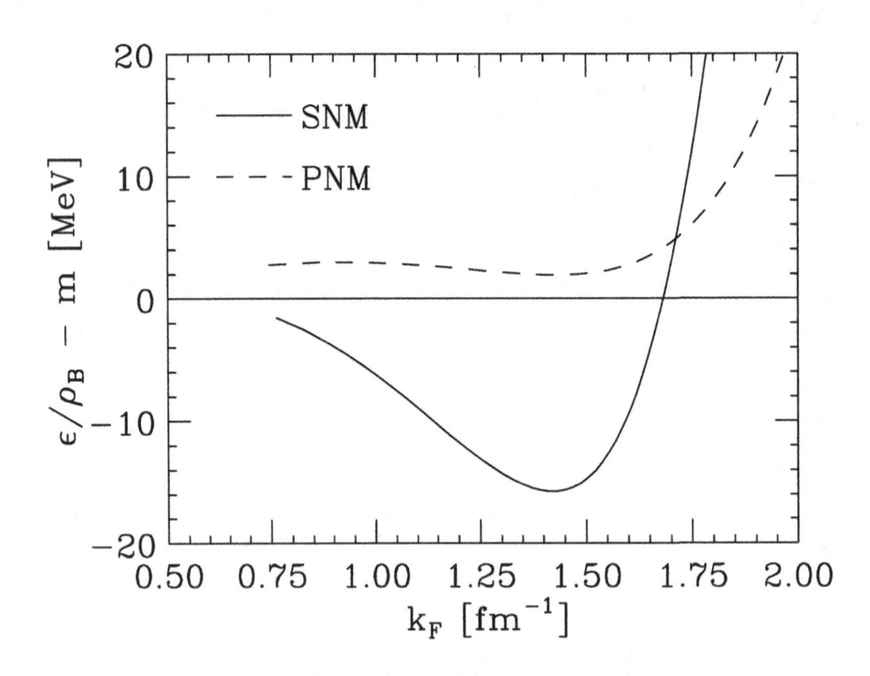

igure 4.28 Fermi momentum dependence of the binding energy per nucleon of SNM (solid line) nd PNM (dashed line) obtained from the Walecka model of Ref. [108] within mean field approxination.

Figure 4.28 shows the binding energies of SNM and PNM obtained from the Walecka nodel within the MFA, plotted against the Fermi momentum k_F.

The Walecka model and its generalisations have been widely employed to study neutron tar properties. However, the relativistic mean field approach, while being very elegant, is ased on a somewhat oversimplified dynamical model, which cannot be constrained using scattering data and is inherently unable to include one-pion-exchange interactions. In ition, while being in principle well suited to describe matter in the $\varrho \to \infty$ limit, its licability at finite densities rests on the assumption that the Compton wavelengths of exchanged mesons is large compared to the typical NN separation distance. The range of dity of this assumption in the case of heavy mesons, such as the ω, may be questionable, needs to be carefully investigated.

CHAPTER 5

ADVANCED VARIATIONAL METHODS

The limitation inherent in variational calculations, arising from the choice of the trial wave function, can be overcome using an approach alternative to the Monte Carlo method discussed in Section 4.6. allowing us to perform perturbative calculations in which the variational results correspond to the 0-th order approximation. This scheme has been applied to carry out accurate calculations of a variety of nuclear matter properties, including the ground-state energy and the two-point Green's function.

5.1 CORRELATED BASIS FUNCTIONS THEORY

The description of Fermi systems based on Correlated Basis Functions (CBF) are a natural extension of the variational approach, in which the trial ground state wave function is written in the form (4.37). The basic tenet of CBF perturbation theory—first developed in the 1960s to study the properties of liquid Helium [52]—is that the correlation operator F of Eqs.(4.38) and (4.39), determined from the minimisation of the ground-state energy, can be used to build the complete set of correlated states [77, 109]

$$|\Psi_n) \equiv \frac{F|\Phi_n\rangle}{\langle\Phi_n|F^\dagger F|\Phi_n\rangle}, \tag{5.1}$$

where $|\Phi_n\rangle$ is a n–particle n–hole state of the non interacting Fermi gas. The notation $|\Psi_n)$ is meant to remind that the correlated states defined by Eq.(5.1) are not orthogonal to one another, that is,

$$(\Psi_n|\Psi_m) = \delta_{nm} + S_{nm} \ , \tag{5.2}$$

with $S_{mm} \neq 0$.

Within the CBF scheme, the expectation values

$$E_n^V = (\Psi_n|H|\Psi_n) \, , \tag{5.3}$$

provide the lowest order approximation to the energy of the state $|\Psi_n)$.

The perturbative expansion is based on the decomposition

$$H = H_0 + H_1 \, . \tag{5.4}$$

where H_0 and H_1 are defined in terms of their matrix elements in the correlated basis, by singling out the diagonal and off-diagonal contributions. The resulting expressions are

$$(\Psi_m|H_0|\Psi_n) = \delta_{mn}(\Psi_m|H|\Psi_n) = \delta_{mn}E_n^V \ , \tag{5.5}$$

$$(\Psi_m|H_1|\Psi_n) = (1-\delta_{mn})\,(\Psi_m|H|\Psi_n) \ . \tag{5.6}$$

If the correlation operator F is properly chosen, so that E_0^V is close to the ground-state energy E_0, the correlated states have large overlaps with the eigenstates of H, and the matrix elements of H_1 are small. As a consequence, a perturbative expansion in powers of H_1 is rapidly convergent, although the use of a non orthogonal basis brings in non-trivial difficulties.

Early applications of the CBF approach exploited the formalism of non-orthogonal perturbation theory to carry out calculations using the states $|\Psi_n)$. Within this scheme, the perturbative corrections to the variational energies E_n^V can be arranged in a series formally identical to that obtained using an orthogonal basis. One finds

$$\begin{aligned}\Delta E_n = E_n - E_n^V &= \sum_{p\neq n} \frac{W_{np}^n W_{pn}^n}{E_n - E_p^V} \\ &+ \sum_{p,q\neq n} \frac{W_{np}^n W_{pq}^n W_{qn}^n}{(E_n - E_p^V)(E_n - E_q^V)} + \dots \ ,\end{aligned} \tag{5.7}$$

where E_n is the n-th eigenvalue of the full Hamiltonian, H, and

$$W_{nm}^n = H_{nm} - E_n S_{mn} \ . \tag{5.8}$$

A notable difference with respect to the case of an orthogonal basis originates from the matrix elements W_{nm}^n, which involve an additional energy dependence besides the one associated with the denominators. Moreover, W^n is a many-body operator, embodying the whole correlation structure induced by the operator F.

The value of the energy shift ΔE_n is found by expanding both W_{np}^n and $(E_n - E_p^V)^{-1}$ around $E_n = E_n^V$, and substituting the resulting expressions

$$W_{np}^n = H_{np} - \Delta E_n S_{np} = H_{np} - E_n^V S_{np} - \Delta E_n S_{np} = \mathcal{W}_{np}^n - \Delta E_n S_{np} \ , \tag{5.9}$$

with $\mathcal{W}_{np}^n = H_{np} - E_n^V S_{np}$, and

$$\frac{1}{E_n - E_p^V} = \frac{1}{E_n^V - E_p^V + \Delta E_n} = \frac{1}{E_n^V - E_p^V} - \frac{1}{E_n^V - E_p^V}\Delta E_n \frac{1}{E_n^V - E_p^V} + \dots \tag{5.10}$$

in the right-hand side of Eq.(5.7). Following this procedure, one arrives at

$$\begin{aligned}\Delta E_n &= \sum_{p\neq n} \frac{\mathcal{W}_{np}^n \mathcal{W}_{pn}^n}{E_n^V - E_p^V} + \dots \\ &= \sum_{p\neq n} \frac{(H_{np} - E_n^V S_{np})(H_{pn} - E_n^V S_{pn})}{E_n^V - E_p^V} + \dots \ .\end{aligned} \tag{5.11}$$

At any given order, ΔE_n exhibits a catastrophic dependence on the particle number. However, as in the case of standard perturbation theory, the divergent terms appearing at different orders cancel against each other. In addition, it has been proved that only linked diagrams contribute to W^n_{nm} and S_{nm}. Therefore, the CBF perturbative series is well behaved in the thermodynamic limit [110].

Enforcing orthogonalisation using the Löwdin transformation [111], which amounts to orthogonalising the correlated n-particle n-hole states among themselves, proves to be problematic, in that the diagonal matrix elements of the Hamiltonian turn out to be larger than the variational estimates. This problem can be circumvented using a two-step orthogonalisation procedure—referred to as Löwdin–Schmidt, or LS, orthogonalisation—that preserves the diagonal matrix elements of the Hamiltonian by construction [112]. Within this scheme, two–particle two–hole states are orthogonal to each other, as well as to all m–particle m–hole states with $m > 2$[1].

In principle, CBF perturbation theory provides a consistent and systematic framework to improve the quality of variational calculations, as well as to perform accurate calculations of nuclear matter properties other than the ground-state energy. However, numerical implementation of the CBF formalism requires the calculation of matrix elements involving correlated states, which entail the non-trivial difficulties discussed in Section 4.4.

Being capable to provide precise evaluations of ground-state expectation values only, the approach based on Monte Carlo techniques appears to be somewhat limited for the scope of CBF perturbation theory. A widely employed alternative method is based on the use of techniques allowing to sum selected classes of terms to all orders of the cluster expansion.

5.2 HYPER-NETTED-CHAIN SUMMATION SCHEME

The failure of low-order approximations of the cluster expansion, illustrated by the examples discussed in Section 4.4.3, indicates that an accurate description of correlations at both short and long range requires the inclusion of many-particle cluster contributions.

The techniques allowing to improve on the low-order approximation—involving the evaluation of infinite sums of terms of the cluster expansion—are based on a classification of the corresponding diagrams according to their topological structures. The first such analysis was performed at the end of the 1950s for the pair distribution function of classical liquids [113]. The starting point is the identification of basic structures, that can be then used to construct more complex diagrams by successive series and parallel connections. The contributions of the resulting diagrams can be summed up to all orders by means of an integral equation, referred to as Hyper-Netted-Chain, or HNC, equation [114].

5.2.1 Fermi Hyper-Netted Chain

The extension of the HNC summation summation method to quantum Bose and Fermi liquids dates back to the 1960s and 1970s [52, 115]. In the case of Fermi systems described using spin and isospin independent Jastrow correlation functions, the result is a system of coupled integral equations, dubbed Fermi Hyper-Netted Chain, or FHNC, equations.

In the following, we will discuss the FHNC equations in the form derived by S. Fantoni and S. Rosati [115], which is best suited to describe strong short-range correlations. It has to be mentioned, however, that a complementary scheme, optimised for the description of systems in which long-range correlations are dominant, has been developed by E. Krotscheck and M. Ristig [116].

We will first consider the simplified case of Jastrow correlation functions, but will not make use of the exact cancellation of cluster contributions corresponding to reducible dia-

[1] Recall that in nuclear mater, owing to translation invariance, the occurrence of one-particle one-hole intermediate states is forbidden by momentum conservation.

grams, which has been rigorously proved for Fermi systems [73]. This scheme will be referred to as Renormalised FHNC, or RFHNC.

Vertex-corrected Irreducible diagrams

Let us start classifying the diagrams associated with the terms of the cluster expansion considering reducible and irreducible topologies.

Consider, for example, the reducible diagram (4.13.a), in which the correlation line $h(\mathbf{r}_{34})$ is attached to vertex 3 of the irreducible diagram (4.9.a). In exactly the same way, one can attach to vertex 3 all possible linked one-body diagrams, that is, diagrams with one open dot representing an active particle. This amounts to multiplying the contribution of graph (4.9.a) by the one-body correlation function $g_1(\mathbf{r}_3)$.

As a result, the sum of all the reducible diagrams having diagram (4.9.a) as the irreducible part, can be represented by a diagram identical to (4.9.a), with vertex 3 renormalised through a multiplication by the vertex correction

$$\xi_d(\mathbf{r}_3) = g_1(\mathbf{r}_3) = 1 + \sum (\text{linked one-body cluster terms}) \ . \tag{5.12}$$

Recall that, owing to translation invariance, in nuclear matter $\xi_d(\mathbf{r})$ is actually a constant, independent of $\mathbf{r}$.

The diagrammatic representation of $g_1(\mathbf{r})$ is given in Fig. (5.1).

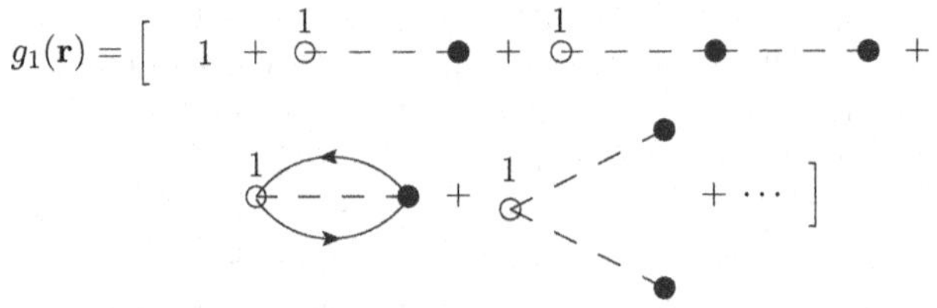

Figure 5.1 Diagrams contributing to $g_1(\mathbf{r}_1) = \xi_d(\mathbf{r}_1)$, see Eq.(5.12).

Note that there are no topological restrictions on the terms to be included in the sum of Eq.(5.12), because in diagram (4.9.a) there are no exchange lines reaching vertex 3. On the other hand, if the reducibility point is reached by exchange lines, like, for example, vertex 3 of diagram (4.9.b), the general diagrammatic rules of Section 4.4.1 prescribe that vertex correction can only by attached through a dashed line, representing a dynamical correlation. Diagrams in which the active point is reached by exchange lines, such as the first diagram in the second line of Fig. 5.1, are not allowed.

From the above discussion, it follows that the description of vertex corrections requires two vertex functions, denoted ξ_d and ξ_e. The former, given by Eq. (5.12) corresponds to d-type vertices, reached by dashed lines only, while the second is attached to e-type vertices, belonging to an exchange loop.

There is one additional vertex function, $\xi_c = \xi_e - 1$, needed to take into account diagrams in which the reducibility point is an internal point connected to the part of the diagram containing the external points through exchange lines only. Note that ξ_c cannot be the same as ξ_e, since any internal point must be reached by at least one correlation line.

In conclusion, the sum of reducible and irreducible diagrams includes all vertex-corrected irreducible, or VIC, diagrams, that is, irreducible diagrams whose vertices are modified by the corrections ξ_d, ξ_e or ξ_c. Note that the second diagrammatic rule of Section 4.4.1 does not apply to VIC diagrams. After correcting with ξ_c, internal points reached by exchange lines only are in fact allowed.

For a translationally invariant system, such as nuclear matter, the one-body distribution function is constant and equal to one. As a consequence, the requirement $\xi_d(\mathbf{r}) = 1$, independent of $\mathbf{r}$, provides a useful benchmark to test the accuracy of calculations of the vertex corrections.

It is convenient to introduce the new sets of one-body VIC diagrams $U_d(\mathbf{r}_1)$ and $U_e(\mathbf{r}_1)$, defined through the equations

$$\xi_d(\mathbf{r}_1) = [1 + U_e(\mathbf{r}_1)] \exp[U_d(\mathbf{r}_1)] \ , \tag{5.13}$$

$$\xi_e(\mathbf{r}_1) = \exp[U_d(\mathbf{r}_1)] \ . \tag{5.14}$$

The external point of the diagrams belonging to $U_d(\mathbf{r}_1)$, labelled 1, is not reached by any exchange lines, while in those belonging to $U_e(\mathbf{r}_1)$ point 1 must be included in a loop of exchange lines. In addition, the diagrams comprised in both $U_d(\mathbf{r}_1)$ and $U_e(\mathbf{r}_1)$ cannot consist of parts connected through point 1. The exponential in the above equations accounts for the fact that any number of d-type structures of $U_d(\mathbf{r}_1)$ can be attached to vertex 1. Because the symmetry factor for n topologically identical structures is $1/n!$, the full contribution from $U_d(\mathbf{r}_1)$ is then given by $\sum_n (1/n!)[U_d(\mathbf{r}_1)]^n$.

For illustration, some of the diagrams included in $U_d(\mathbf{r}_1)$ and to $U_e(\mathbf{r}_1)$ are shown in Fig. 5.2.

Simple and composite diagrams

An i–j subdiagram is a part of an irreducible diagram that is only connected to the rest through the two points labelled i and j [113].

Irreducible diagrams are classified as either simple, or composite. Simple diagrams consist of only one 1–2 subdiagram, where the labels 1 and 2 correspond to the external points. Composite, or X-type, diagrams, on the other hand, contain two or more 1–2 subdiagrams, which are said to be connected in parallel through points 1 and 2. As no integration is performed over the coordinates of the external points, the contributions of composite diagrams factorise.

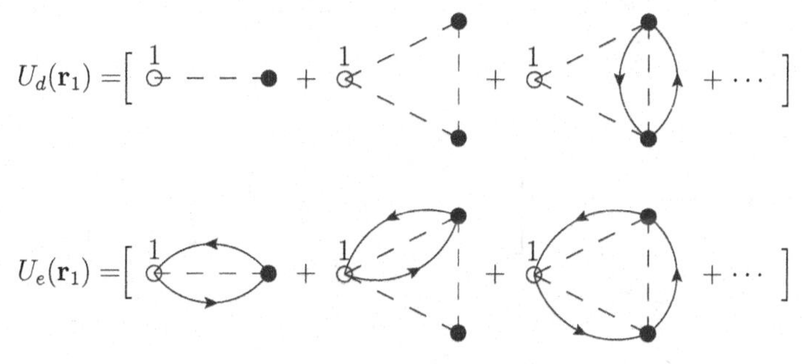

$U_d(\mathbf{r}_1) = [\ \ldots + \cdots\ ]$

$U_e(\mathbf{r}_1) = [\ \ldots + \cdots\ ]$

Figure 5.2 Examples of diagrams belonging to $U_d(\mathbf{r}_1)$ and to $U_e(\mathbf{r}_1)$.

Diagram (4.9.a) and the diagrams of Figs. 4.11 and 4.12 are examples of simple diagrams. A composite diagram, with the factorisation of the associated subdiagrams, is is illustrated in Fig. 5.3.

Composite diagrams can be further classified according to the properties of the external points 1 and 2. The set of diagrams with only dynamical correlation lines reaching points 1 and 2, like diagram (5.4.a), is denoted X_{dd}. When two exchange lines are attached to point 1 or 2 the corresponding composite set of diagrams are denoted X_{ed} or X_{de}, see diagrams (5.4.b) and (5.4.c). Likewise, diagram (5.4.d) belongs to the set X_{ee}, since two statistical lines arrive at both external points. To build *ee* diagrams with the external points in the same exchange loop, it is convenient to define composite diagrams with an open loop of exchange lines starting from the external point 1 and ending at the external point 2, like, for instance, diagram (5.4.e) and the one shown in Fig. 5.3. Note that, in the case of Jastrow correlation functions, the set of these diagrams, denoted by X_{cc}, does not contribute directly to the pair distribution function.

Nodal diagrams

Simple diagrams can be also subdivided into two sets: nodal, or chain, diagrams and elementary diagrams. A diagram is called nodal, or N-type, if it has at least one node, that is, a point such that all paths of lines going from one external point to the other pass through it. Diagrams (4.9.a), (4.11.a) and (4.11.b) are nodal. An elementary, or E-type, diagram is a simple diagram that is not nodal, see, for example, diagram (4.12.a).

Nodal diagrams can be classified according to the properties of the external points, following the same scheme employed for composite diagrams. Examples of diagrams belonging to the sets N_{dd}, N_{ed}, N_{de}, N_{ee} and N_{cc} are displayed in Fig. 5.5.

5.2.2 RFHNC equations

To introduce the method of derivation of the RFHNC equations, let us consider a nodal diagram contributing to the function $N_{xy}(r_{12})$, and assign the label 3 to the node closest to vertex 1. The contribution of all nodal diagrams can be obtained by convoluting the sum of all non-nodal 1–3 subdiagrams, $X_{xx'}(r_{13})$, with the full set of 3–2 subdiagrams, with or without nodes, $X_{y'y}(r_{23}) + N_{y'y}(r_{23})$. The resulting integral equation reads

$$N_{xy}(r_{12}) = \sum_{x'y'} \varrho \int d\mathbf{r}_3 \; X_{xx'}(r_{13}) \zeta_{x'y'} \left[X_{y'y}(r_{23}) + N_{y'y}(r_{23}) \right] , \tag{5.15}$$

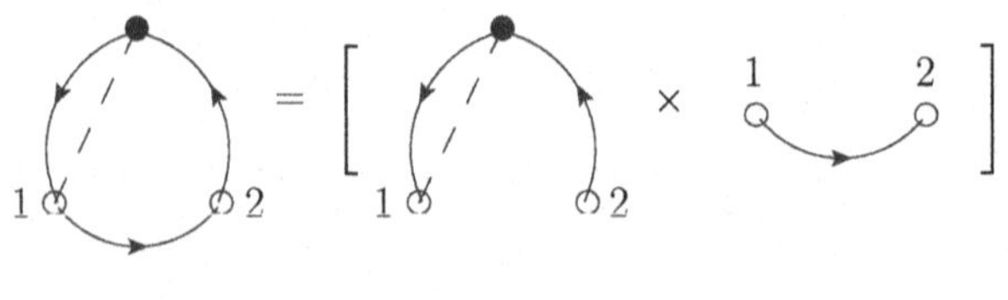

Figure 5.3 Example of a composite diagram factorised into two simple subdiagrams.

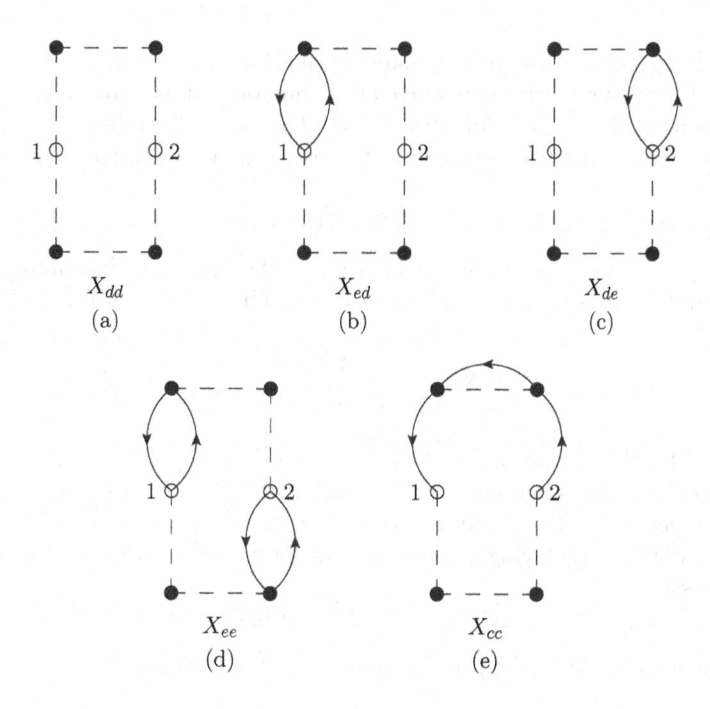

Figure 5.4 Classification of composite diagrams based on the properties of the external points.

where the indices x, y, x', and y'specify the type—d or e—of external points. The coefficients $\zeta_{x'y'}$, account for vertex corrections and the proper treatment of exchange loops. For the possible $x'y'$ combinations, dd, de, and ed, their values are

$$\zeta_{dd} = \xi_d \quad , \quad \zeta_{de} = \zeta_{ed} = \xi_e \ , \tag{5.16}$$

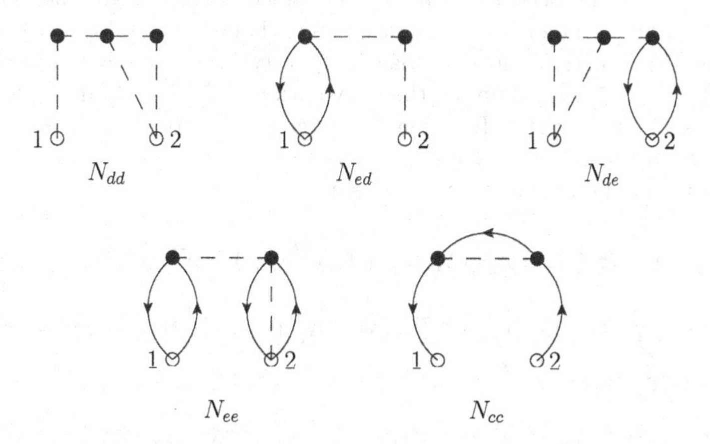

Figure 5.5 Classification of nodal diagrams according to the properties of the external points.

whereas $\zeta_{ee} = 0$.

In order to highlight the role of the chain summation of nodal diagrams in determining the long-range behaviour of the pair correlation function, it is convenient to rewrite the convolution equation (5.15) in momentum space. Omitting the subscripts referring to the types of vertices, which are not relevant in this context, the result can be written in the form

$$\tilde{N}(k) = \varrho\, \xi\, \tilde{X}(k)[\tilde{X}(k) + \tilde{N}(k)] \, , \tag{5.17}$$

where $\tilde{N}(k)$ and $\tilde{X}(k)$ are the Fourier transforms of the functions describing the contributions of nodal and composite diagrams, respectively. The solution of Eq. (5.17) is

$$\tilde{N}(k) = \frac{\varrho\, \xi \tilde{X}(k)}{1 - \varrho \xi \tilde{X}(k)} \, . \tag{5.18}$$

Let us now consider the simple case of a nodal diagram $N^{(n)}(r)$, consisting of n dynamical correlation lines $h(r)$. For example, $N^{(2)}(r)$ and $N^{(3)}(r)$ correspond to diagrams (4.9.a) and (4.11.a), respectively. The analytic expression of $\tilde{N}^{(n)}(k)$ can be easily derived from Eq. (5.17) by iteration, replacing $\tilde{X}(k)$ with its lowest order approximation, $\tilde{h}(k)$. The resulting exapression

$$\tilde{N}^{(n)}(k) = (\varrho\, \xi)^{n-1}\, \tilde{h}(k)^n \, , \tag{5.19}$$

implies that the sum of all $N^{(n)}(k)$, to be compared to Eq. (5.18), is

$$\sum_n \tilde{N}^n(k) = \frac{\varrho\, \xi \tilde{h}^2(k)}{1 - \varrho \xi \tilde{h}(k)} \, . \tag{5.20}$$

If $h(r)$ is long ranged, say $h(r) \to \alpha/r^2$—with α a constant—as $r \to \infty$, then in the long wavelength limit, corresponding to $k \to 0$, $\tilde{h}(k) \to -\alpha/(2\pi k)$. Hence, in this limit the functions $\tilde{N}^{(n)}(k)$ turn out to diverge more strongly than $\tilde{h}(k)$. On the other hand, their sum is well behaved, being only as divergent as $\tilde{h}(k)$. This result shows that, while the cluster expansion in the number of points diverges at any finite order, the result of the chain summation exhibits the correct long-range behaviour.

The equation for N_{cc}, cannot be written in the same form as Eq. (5.15). In this case, the contributing diagrams need to be further classified according to whether the external points are reached by exchange lines only or by an exchange line and at least one dynamical correlation line. This distinction gives rise to four types of nodal cc functions, denoted N_{cc}^{hh}, $N_{cc}^{h\ell}$, $N_{cc}^{\ell h}$ and $N_{cc}^{\ell\ell}$. For example, the convolution of N_{cc}^{hh} with any of the other cyclic functions brings about a vertex function ξ_e, whereas the convolution of two $N_{cc}^{\ell\ell}$ cyclic functions requires a vertex correction ξ_c.

The equations for the four nodal functions are

$$N_{cc}^{hh}(r_{12}) = \varrho \int d\mathbf{r}_3 \xi_e X_{cc}(r_{13})[X_{cc}(r_{32}) + N_{cc}^{hh}(r_{32}) + N_{cc}^{\ell h}(r_{32})] \, , \tag{5.21}$$

$$N_{cc}^{h\ell}(r_{12}) = \varrho \int d\mathbf{r}_3 \xi_e X_{cc}(r_{13}) \left[N_{cc}^{h\ell}(r_{32}) + N_{cc}^{\ell\ell}(r_{32}) - \frac{1}{\nu}\ell(r_{32}) \right] , \tag{5.22}$$

$$N_{cc}^{\ell h}(r_{12}) = N_{cc}^{h\ell}(r_{12}) \, , \tag{5.23}$$

$$N_{cc}^{\ell\ell}(r_{12}) = -\frac{\varrho}{\nu} \int d\mathbf{r}_3 \left\{ \xi_e \ell(r_{13}) N_{cc}^{h\ell}(r_{32}) + \xi_c \ell(r_{13}) \left[N_{cc}^{\ell\ell}(r_{32}) - \frac{1}{\nu}\ell(r_{32}) \right] \right\} , \tag{5.24}$$

with $N_{cc}^{\ell h}(r_{12}) = N_{cc}^{h\ell}(r_{12})$.

The total cc nodal function, given by

$$N_{cc}(r_{12}) = N_{cc}^{hh}(r_{12}) + N_{cc}^{h\ell}(r_{12}) + N_{cc}^{\ell h}(r_{12}) + N_{cc}^{\ell\ell}(r_{12}) \ , \tag{5.25}$$

is the solution of the integral equations obtained from the sum of Eqs. (5.21)–(5.24)

$$\begin{aligned} N_{cc}(r_{12}) = \varrho \int d\mathbf{r}_3 X_{cc}(r_{13})\xi_e \Big[X_{cc}(r_{32}) + N_{cc}(r_{32}) - \frac{\ell(r_{32})}{\nu}\Big] \\ - \frac{\ell(r_{13})}{\nu}\xi_e \Big[X_{cc}(r_{32}) + \mathcal{P}(r_{32})\Big] \\ - \frac{\ell(r_{13})}{\nu}\xi_c \Big[-\frac{\ell(r_{32})}{\nu} + N_{cc}(r_{32}) - \mathcal{P}(r_{32})\Big] \ , \end{aligned} \tag{5.26}$$

with $\mathcal{P}$ defined as

$$\mathcal{P}(r_{12}) = \rho \int d\mathbf{r}_3 \xi_e X_{cc}(r_{13}) \Big[X_{cc}(r_{32}) + N_{cc}(r_{32}) - \frac{\ell(r_{32})}{\nu}\Big] \ . \tag{5.27}$$

The contributions to the pair distribution function can also be classified according to the properties of the external points, and the resulting functions can in turn be expressed in terms of the contributions arising from simple and composite diagrams. Collecting the corresponding expressions, referred to as partial distribution functions, we obtain

$$g_2(r_{12}) = \xi_d^2 g_{dd}(r_{12}) + \xi_d\xi_e[g_{de}(r_{12}) + g_{ed}(r_{12})] + \xi_e^2 g_{ee}(r_{12}) \ , \tag{5.28}$$

where

$$\begin{aligned} g_{dd}(r_{12}) &= f^2(r_{12})\exp[N_{dd}(r_{12}) + E_{dd}(r_{12})] \ , \\ g_{de}(r_{12}) &= g_{ed}(r_{12}) = N_{de}(r_{12}) + X_{de}(r_{12}) \ , \\ g_{ee}(r_{12}) &= g_{dd}(r_{12})\Big\{N_{ee}(r_{12}) + E_{ee}(r_{12}) + [N_{de}(r_{12}) + E_{de}(r_{12})]^2 \\ &\quad - \nu\Big[N_{cc}(r_{12}) - \frac{1}{d}\ell(r_{12}) + E_{cc}(r_{12})\Big]^2\Big\} \ , \\ g_{cc}(r_{12}) &= N_{cc}(r_{12}) + X_{cc}(r_{12}) - \frac{1}{\nu}\ell(r_{12}) \ . \end{aligned} \tag{5.29}$$

and the functions $E_{xy}(r_{12})$ denote the sums of contributions arising from xy elementary diagrams.

The composite functions, which can be seen as generalised links, are given by

$$\begin{aligned} X_{dd}(r_{12}) &= g_{dd}(r_{12}) - N_{dd}(r_{12}) - 1 \ , \\ X_{de}(r_{12}) &= X_{ed}(r_{12}) = g_{dd}(r_{12})[N_{de}(r_{12}) + E_{de}(r_{12})] - N_{de}(r_{12}) \ , \\ X_{ee}(r_{12}) &= g_{dd}(r_{12})\Big\{N_{ee}(r_{12}) + E_{ee}(r_{12}) + [N_{de}(r_{12}) + E_{de}(r_{12})]^2 \\ &\quad - \nu\Big[N_{cc}(r_{12}) - \frac{1}{\nu}\ell(r_{12}) + E_{cc}(r_{12})\Big]^2\Big\} - N_{ee}(r_{12}) \ , \\ X_{cc}(r_{12}) &= g_{dd}(r_{12})[N_{cc}(r_{12}) - \frac{1}{\nu}\ell(r_{12}) + E_{cc}(r_{12})] - N_{cc}(r_{12}) + \frac{1}{\nu}\ell(r_{12}) \ . \end{aligned} \tag{5.30}$$

Finally, the functions U_d and U_e, entering the definitions of the vertex functions ξ_d and ξ_e of Eqs. (5.13) and (5.14), are obtained from the integral equations

$$\begin{aligned} U_d = \varrho \int d\mathbf{r}_2 \Big\{\xi_d[X_{dd}(r_{12}) - E_{dd}(r_{12}) - S_{dd}(r_{12})(g_{dd}(r_{12}) - 1)] \\ + \xi_e[X_{de}(r_{12}) - E_{de}(r_{12}) - S_{dd}(r_{12})g_{de}(r_{12}) - S_{de}(r_{12})(g_{dd}(r_{12}) - 1)]\Big\} + E_d \ , \end{aligned} \tag{5.31}$$

and

$$U_e = \varrho \int d\mathbf{r}_2 \Big\{ \xi_d [X_{ed}(r_{12}) - E_{ed}(r_{12})] + \xi_e [X_{ee}(r_{12}) - E_{ee}(r_{12})]$$
$$-\xi_d \left[S_{dd}(r_{12}) g_{ed}(r_{12}) + S_{ed}(r_{12})(g_{dd}(r_{12}) - 1)\right] - \xi_e [S_{ee}(r_{12})(g_{dd}(r_{12}) - 1)$$
$$+S_{ed}(r_{12}) g_{de}(r_{12}) + S_{dd}(r_{12}) g_{ee}(r_{12}) + S_{de}(r_{12}) g_{ed}(r_{12}) - 2d S_{cc}(r_{12}) g_{cc}(r_{12})]$$
$$-\ell(r_{12})[\mathcal{N}^{\ell}_{cc}(r_{12}) - \frac{1}{\nu}\ell(r_{12})] \Big\} + E_e \ . \tag{5.32}$$

Here E_d and E_e denote the contributions of one–body elementary diagrams with external points of type d and e, respectively,

$$S_{xy}(\vec{r}_{12}) = \frac{1}{2} N_{xy}(r_{12}) + E_{xy}(r_{12}) \ ,$$

and

$$\mathcal{N}^{\ell}_{cc}(r_{12}) = N^{\ell\ell}_{cc}(r_{12}) + N^{\ell h}_{cc}(r_{12}) \ . \tag{5.33}$$

The RFHNC equations are solved numerically by iterations. For dense systems like liquid helium or nuclear matter, however, achieving convergence is difficult, and often requires the use of an algorithm to smooth out oscillations occurring in the iterative process.

An additional equation for the elementary diagrams would be needed to make the RFHNC system self-contained. However, because of their topological structure, a consistent treatment of the elementary diagrams based on an integral equations with a two-body kernel—similar to those determining the contributions of nodal and composite diagrams—is not possible. Few examples of four-point elementary diagrams are given in Fig. 5.6,

The simplest approximation consists of neglecting elementary diagrams altogether, that is, of setting $E_{xy} = 0$. In nuclear matter, this scheme, referred to as RFHNC/0, provides a good description of the long-range behaviour of the pair distribution function, and it is still fairly accurate at short distances.

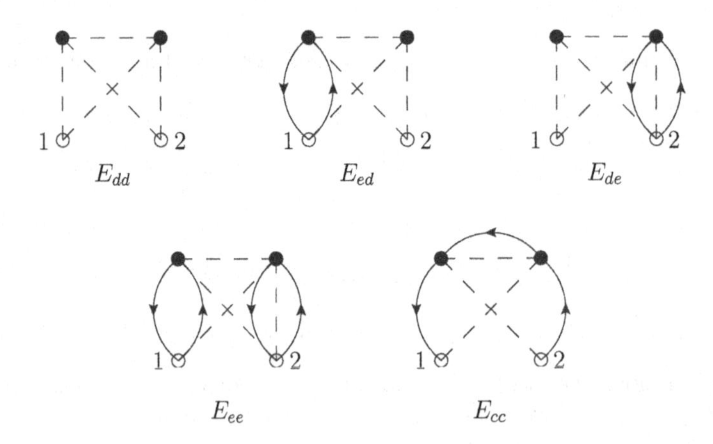

Figure 5.6 Examples of four-point elementary diagrams.

A straightforward approach to include the contributions of elementary diagrams is based on the direct evaluation of the n–body term, E_n, which is then plugged into the RFHNC equations. The corresponding approximation is referred to as RFHNC/n.

The main problem associated with the neglect, or the approximate treatment, of elementary diagrams is the possible violation of the variational principle, which may become significant at high density. The accuracy of RFNHC calculations can be tested comparing their results to those obtained from the Monte Carlo approach described in Section 4.6, as well as studying nuclear matter properties other than the ground-state energy, such as the pair distribution function sum rule of Eq.(4.50).

5.3 EXTENSION TO SPIN-ISOSPIN DEPENDENT CORRELATIONS

In order to obtain accurate upper bounds to the ground-state energy of nuclear matter using the Hamiltonians discussed in Chapter 2, the form of the correlation function F_{ij} of Eq.(4.38) must be chosen is such a way as to reflect the operator structure of the NN potential, see Eqs. (2.12) and (2.13). However, the occurrence of additional contributions arising from non vanishing commutators of the operators O^p severely hampers the generalisation of the RFHNC summation scheme to diagrams whose lines represent spin–isospin dependent correlation function. An approximation designed to alleviate this problem, referred to as Single Operator Chain (RFHNC/SOC) summation scheme, was proposed in the late 1970s [56]. The RFHNC/SOC formalism has been successfully applied to nuclear matter, and extended to the treatment of atomic nuclei with $A \leq 40$ in the 2000s [117].

The expectation value of any two-body operator of the form (2.12) can be conveniently written introducing spin–isospin dependent generalisations of the pair distribution function, defined as

$$g^p(\mathbf{r}_{12}) = \frac{A(A-1)}{\varrho^2} \, \frac{\text{Tr}_1\text{Tr}_2 \int dx_3 \dots dx_A \Psi_T^\dagger(X) F^\dagger O^p_{12} F \Psi_T(X)}{\int dx_1 \dots dx_A \ \Psi_T^\dagger(X) F^\dagger F \Psi_T(X)} = \frac{\mathcal{N}^p}{\mathcal{D}} \ , \tag{5.34}$$

with the operators O^p defined in Section 2.3.1. It is apparent that for $p = 1$ the above equation reduces to the definition of the pair distribution function $g_2(r_{12})$ of Eq.(5.28).

The expectation value of the NN potential reads

$$\langle v \rangle = (0| \sum_{i<j} v_{ij} |0) = \frac{A(A-1)}{2} (0|v_{12}|0) = A \ \frac{\rho}{2} \sum_p \int d\mathbf{r}_{12} \, g^p(\mathbf{r}_{12}) v^p(r_{12}) \ , \tag{5.35}$$

where, according to the philosophy of CBF perturbation theory, $|0)$ is the variational ground state, described by the wave function Ψ_T. Owing to translation invariance, the state dependent distribution functions, like the scalar one, only depend on the magnitude of the relative distance. Hence, the expectation value of the two-body potential diverges with number of particles, while $\langle v \rangle / A$ is a finite quantity.

The numerator of Eq. (5.34), $\mathcal{N}^p$, involves the product

$$F^\dagger O^p_{12} F = X^{(2)}(x_1, x_2) + \sum_{i \neq 1,2} X^{(3)}(x_1, x_2; x_i) + \sum_{i<j \neq 1,2} X^{(4)}(x_1, x_2; x_i, x_j) + \dots . \tag{5.36}$$

where the cluster terms X^n are now n-body operators. For example

$$\begin{aligned} X^{(2)}(x_1, x_2) &= F^\dagger_{12} O^p_{12} F_{12} \\ X^{(2)}(x_1, x_2; x_i) &= (\mathcal{S} F^\dagger_{12} F^\dagger_{1i} F^\dagger_{2i}) O^p_{12} (\mathcal{S} F_{12} F_{1i} F_{2i}) - F^\dagger_{12} O^p F_{12} \ , \end{aligned} \tag{5.37}$$

where $\mathcal{S}$ symmetrises the product appearing on its right. The expansion of $\mathcal{N}^p$ leads to an expression similar to the Eq. (4.61)

$$\mathcal{N}^p = \sum_{n=2}^{A} \frac{\varrho^{n-2}}{(n-2)!} \mathrm{Tr}_1 \mathrm{Tr}_2 \int dx_3 \ldots dx_N X^{(N)}(x_1, x_2; x_3, \ldots, x_N) g_N^{MF}(x_1, \ldots, x_N) \ , \quad (5.38)$$

where the generalised n-body Fermi gas distribution function is given by

$$g_n^{FG}(x_1, \ldots, x_N) = \frac{A!}{(A-N)!} \frac{1}{\varrho^N} \int dx_1 \ldots dx_A \Phi_0(X)^\dagger \Phi_0(X) \ . \quad (5.39)$$

The above definition can be rewritten in the form

$$g_n^{FG}(x_1, \ldots, x_n) = \frac{1}{\varrho^N} \sum_{\mathbf{k}_1 < \cdots < \mathbf{k}_n} \phi^\dagger_{\mathbf{k}_1}(x_1) \ldots \phi^\dagger_{\mathbf{k}_n}(x_n) \mathcal{A}[\phi_{\mathbf{k}_1}(x_1) \ldots \phi_{\mathbf{k}_n}(x_n)] \ , \quad (5.40)$$

where the explicit expression of the antisymmetrisation operator is

$$\mathcal{A} = 1 - \sum_{i<j} P_{ij} + \sum_{i<j<k} (P_{ij}P_{jk} + P_{ik}P_{kj}) + \ldots \ , \quad (5.41)$$

with P_{ij} being the two-body exchange operator in spin-isospin space

$$P_{ij} = \frac{1}{4}(1 + \boldsymbol{\sigma}_i \cdot \boldsymbol{\sigma}_j)(1 + \boldsymbol{\tau}_i \cdot \boldsymbol{\tau}_j) \ . \quad (5.42)$$

By writing the antisymmetrisation operator as in Eq. (5.41), and carrying out the momentum integrations, one finds

$$g_n^{FG}(x_1, \ldots, x_n) = \frac{1}{\nu^n} \sum_{\sigma_i, \tau_i} \chi^\dagger_{\sigma_1} \cdots \chi^\dagger_{\sigma_n} \eta^\dagger_{\tau_1} \cdots \eta^\dagger_{\tau_n} \Big[1 - \sum_{i<j} P_{ij} \ell^2(r_{ij}) \quad (5.43)$$
$$+ \sum_{i<j<k} (P_{ij}P_{jk} + P_{ij}P_{ik}) \ell(r_{ij}) \ell(r_{jk}) \ell(r_{ki}) - \ldots \Big] \chi_{\sigma_1} \cdots \chi_{\sigma_n} \eta_{\tau_1} \cdots \eta_{\tau_n} .$$

This expression is very similar to Eq. (4.62), the only difference being the absence of the traces in spin-isospin space.

Substitution of the above result into Eq. (5.38) yields

$$\mathcal{N}^p = \sum_{n=2}^{A} \frac{\varrho^{n-2}}{(N-2)!} \int d\mathbf{r}_3 \ldots, \mathbf{r}_n \ \mathrm{CTr}_1 \ldots \mathrm{CTr}_n \Big\{ X^{(n)}(x_1, x_2; x_3, \ldots, x_N) \Big[1 - \sum_{i<j} P_{ij} \ell^2(r_{ij})$$
$$+ \sum_{i<j<k} (P_{ij}P_{jk} + P_{ij}P_{ik}) \ell(r_{ij}) \ell(r_{jk}) \ell(r_{ki}) - \ldots \Big] \Big\} \ , \quad (5.44)$$

where the symbol CTr denotes the trace normalised in such a way as to have $\mathrm{CTr}(\mathbf{1}) = 1$

The important property stated by Eq. (4.63), leading to the cancellation between the unlinked diagrams of the numerator and the denominator, still holds true for the generalised for g_n^{FG}.

5.3.1 Diagrammatic rules

In addition to the dashed lines representing the function $h(r_{ij}) = f^c(r_{ij})^2 - 1$, the RFHNC/SOC diagrams involve single and double wavy lines associated with the functions[2]

$$2f^c(r_{ij})f^{p>1}(r_{ij}) \quad , \quad f^{p>1}(r_{ij})f^{q>1}(r_{ij}) \ , \tag{5.45}$$

where the factor 2 accounts for the fact that in the product $F^\dagger O^p_{12} F$ the central correlation function appears on both sides of O^p_{12}. Wavy lines are labelled with the index of the $p > 1$ component of the correlation function, see Fig. 5.7.

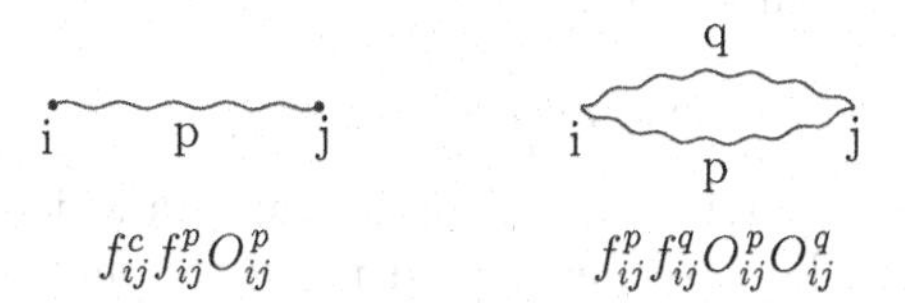

Figure 5.7 Diagrammatic representation of products involving the correlation functions $f^{p>1}$.

In addition, thick solid lines, as the one shown in Fig. 5.8, are introduced to represent the terms contributing to the product $F_{12}v_{12}F_{12}$, Fig. 5.8

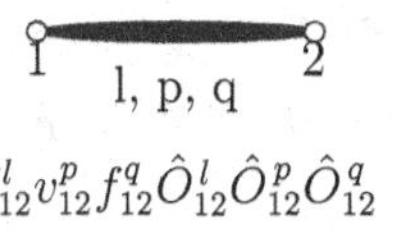

1 2

l, p, q

$f^l_{12} v^p_{12} f^q_{12} \hat{O}^l_{12} \hat{O}^p_{12} \hat{O}^q_{12}$

Figure 5.8 Diagrammatic representation of the contributions to the product $F_{12}v_{12}F_{12}$.

Note that, because the contributions of the diagrams in general depends on the order of the operators involved, all permutations need to be considered.

The diagrammatic classification is similar to the one discussed in Chapter 4. As mentioned above, disconnected diagrams of the numerator exactly cancel with the denominator, and only connected diagrams must be taken into account. On the other hand, unlike the case of Jastrow correlations, reducible diagrams do not exactly cancel. The details of the treatment of reducible diagrams can be found in Appendix A.

Equation (5.38) shows that calculation of $\langle v \rangle / A$ involves the traces of spin–isospin dependent operators entering the definition of both the potential and the correlation function. They are evaluated using Pauli's identity

$$(\mathbf{a} \cdot \boldsymbol{\sigma})(\mathbf{b} \cdot \boldsymbol{\sigma}) = (\mathbf{a} \cdot \mathbf{b}) + i\, \boldsymbol{\sigma} \cdot (\mathbf{a} \times \mathbf{b}) \ , \tag{5.46}$$

with $\mathbf{a}$ and $\mathbf{b}$ being any vectors independent of $\boldsymbol{\sigma}$. The above equation, which also applies to Pauli matrices in isospin space, allows us to rewrite a generic operator product as

$$\prod O_{ij} = C + \text{rest} \ , \tag{5.47}$$

[2]In the following, we will also use the alternative notation $p = 1, \ldots, 6 \to c,\ \tau,\ \sigma,\ \sigma\tau,\ t,\ t\tau$.

where C does not contain any spin–isospin dependent operators, while the rest contains terms at least linear in $\boldsymbol{\sigma}$ and $\boldsymbol{\tau}$. Because

$$\text{CTr}\ \boldsymbol{\sigma} = \text{CTr}\ \boldsymbol{\tau} = 0\ , \tag{5.48}$$

the only non vanishing contribution to $\prod O_{ij}$ is C, which, in general, depends on the order of the operators involved in the product.

The operator structures appearing in the diagrams are divided in following three classes.

Products of operators acting on the same pair

Let us consider a cluster term in which the points i and j are joined by two operators. The above discussion implies

$$\text{CTr}\ O^p_{ij} O^q_{ij} = A^p \delta_{pq}\ , \tag{5.49}$$

with $A^p = 1, 3, 3, 9, 6, 18$ for $p = 1, 6$. The CTr of diagrams in which more than two operators act on the same pair ij can be easily evaluated using the matrices K^{pqr}, defined by the equation

$$O^p_{ij} O^q_{ij} = \sum_r K^{pqr} O^r_{ij}. \tag{5.50}$$

Comparing the last two equations it is readily seen that $K^{pq1} = \delta_{pq} A_p$. Using Eq. (5.50) it turns out that

$$\text{CTr}(O^p_{ij} O^q_{ij} O^r_{ij}) = \sum_s K^{pqs} \text{CTr}(O^s_{ij} O^r_{ij}) = K^{pqr} A^r\ . \tag{5.51}$$

Note that, since operators acting on the same pair of points commute, the order of operator in the previous equation is immaterial, thus

$$K^{pqr} A^r = K^{qpr} A^r = K^{qrp} A^p \ \ldots \tag{5.52}$$

Single Operator Rings (SOR)

Single operator rings, like the one displayed in Fig 5.9, are characterised by having at most two operators with a common index. The normalized trace of a SOR does not depend on the ordering of the operators having the point i in common. Owing to Pauli's identity, the non commuting terms are in fact linear in either $\boldsymbol{\sigma}_i$ or $\boldsymbol{\tau}_i$, and their traces vanish.

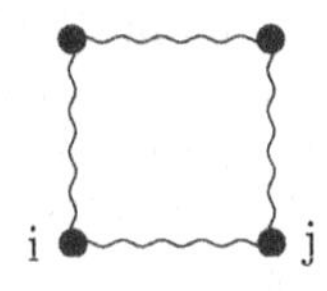

Figure 5.9 Four-body SOR diagram.

Let O^p_{ij} and O^q_{jk} be the only two operators arriving at point j. Making use again of Pauli's identity it is possible to completely eliminate the operatorial dependence on point j. Integrating over the azimuthal angle ϕ_j and tracing over the spin-isospin degrees of freedom of particle j yields

$$\text{CTr}_j \int d\phi_j O^p_{ij} O^q_{jk} = \sum_r \int d\phi_j \xi^{pqr}_{ijk} O^r_{ij}\ , \tag{5.53}$$

where the coefficients ξ^{pqr}_{ink} depend on the internal angles of the triangle formed by the vectors $\mathbf{r}_{ij}$, $\mathbf{r}_{jk}$, $\mathbf{r}_{ik}$

$$
\begin{aligned}
\xi^{\sigma\sigma r}_{ijk} &= \delta_{\sigma r} \ , \\
\xi^{\sigma tr}_{ijk} &= \delta_{tr}\frac{1}{2}(3\cos^2\theta_k - 1) \ , \\
\xi^{t\sigma r}_{ijk} &= \delta_{tr}\frac{1}{2}(3\cos^2\theta_i - 1) \ , \\
\xi^{ttr}_{ijk} &= \delta_{\sigma r}(3\cos^2\theta_j - 1) \\
&+ \delta_{tr}\frac{1}{2}[-9\cos\theta_i\cos\theta_j\cos\theta_k - 3(\cos^2\theta_i + \cos^2\theta_j + \cos^2\theta_k) + 2] \ , \\
\xi^{\tau\tau r}_{ijk} &= \delta_{\tau r} \ , \\
\xi^{(p\tau)(q\tau)(r\tau)}_{ijk} &= \xi^{pqr}_{ijk} \ ,
\end{aligned}
\tag{5.54}
$$

with $p, q, r = \sigma, t$. The evaluation of SOR diagrams is quite simple. The operators with one common point are placed next to each other, and Eq. (5.53) is exploited to remove the dependence on the coordinates of the common points.

Multiple operator diagrams

Consider the normalized trace of diagram (a) of Fig. 5.10, where more than two operators arrive at both points i and j. In principle, all possible orderings of the operators have to be considered in the evaluation of the normalized trace. However, invariance under cyclic permutations is a general property of the traces. As a consequence, there are only two different orderings of the operators: a "successive" order, in which O^p_{ij} and O^q_{ij} can be placed next to each other, and an "alternate" order, in wich either O^r_{ik} or O^s_{jk} is placed between them. For the successive order, using Eq. (5.50), (5.51) and (5.53) we obtain

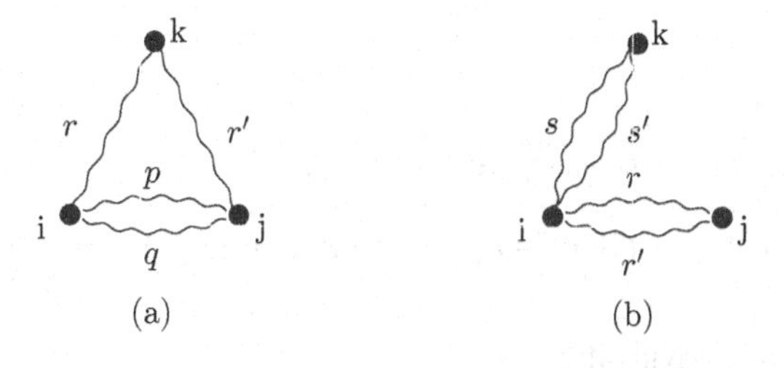

Figure 5.10 Examples of multiple operator diagram

$$
\begin{aligned}
\int d\phi_i \mathrm{CTr}\; O^p_{ij} O^q_{ij} O^{r'}_{ik} O^{r''}_{jk} &= \sum_r K^{pqr} \int d\phi_k \mathrm{CTr}\; O^r_{ij} O^{r'}_{ik} O^{r''}_{jk} \\
&= \sum_{r,r'''} K^{pqr} \int d\phi_k \xi^{r'r''r'''}_{ikj} \mathrm{CTr}\; O^r_{ij} O^{r'''}_{ij} \\
&= \sum_r K^{pqr} A^r \int d\phi_k \xi^{r'r''r}_{ikj} \ .
\end{aligned}
\tag{5.55}
$$

On the other hand, for the alternate order we find

$$\int d\phi_i \mathrm{CTr} O^p_{ij} O^{r'}_{ik} O^q_{ij} O^{r''}_{jk} = \sum_r L^{pqr} \int d\phi_k \xi^{r'r''r}_{ikj} . \tag{5.56}$$

To determine the matrix L^{pqr} we note that either

$$\mathrm{CTr}\ O^p_{ij} [O^{r'}_{ik}, O^q_{ij}] O^{r''}_{jk} = 0 \tag{5.57}$$

or

$$\mathrm{CTr}\ O^p_{ij} \{O^{r'}_{ik}, O^q_{ij}\} O^{r''}_{jk} = 0 . \tag{5.58}$$

It can be easily seen that in the first and second case the results are $L^{pqr} = K^{pqr} A^r$ and $L^{pqr} = -K^{pqr} A^r$, respectively.

Another possibility that needs to be discussed involves two SOR meeting at point i, like in diagram (b) of Fig. 5.10. Again, owing to the invariance of the trace upon cyclic exchanges, there are only two distinct cases. When the two operators acting on the pairs ij and ik are next to each other, it turns out that

$$\mathrm{CTr}\ O^r_{ij} O^{r'}_{ij} O^s_{ik} O^{s'}_{ik} = \delta_{rr'} A^r \delta_{ss'} A^s , \tag{5.59}$$

where we have used $K^{pq1} = \delta_{pq} A_p$.

In order to deal with the alternate order, we introduce the matrix D_{rs}, defined through the equation

$$\sum_{\vec{\sigma}_i \vec{\tau}_i} O^r_{ij} O^s_{ik} O^{r'}_{ij} = \delta_{rr'} A^r (1 + D_{rs}) O^s_{ik} , \tag{5.60}$$

which in the case of tensor operators implies an integration over the azimuthal angle ϕ. The values of D_{rs}, that depend on the type of the operators O^r and O^s, turn out to be

$$D_{\sigma\tau} = 0 \ , \quad D_{(\sigma\tau, t\tau)(\sigma\tau, t\tau)} = -\frac{8}{9} ,$$
$$D_{\sigma\sigma} = D_{\tau\tau} = D_{(\sigma,\tau)(\sigma\tau, t\tau)} - \frac{4}{3} . \tag{5.61}$$

Hence, for the trace in alternate order we find the result

$$\mathrm{CTr}\ O^r_{ij} O^s_{ik} O^{r'}_{ij} O^{s'}_{ik} = \delta_{rr'} A^r (1 + D_{rs}) \delta_{ss'} A^s . \tag{5.62}$$

5.3.2 RFHNC/SOC approximation

In the case of correlations involving non-commuting operators, the full RFHNC summation of linked cluster diagrams involves prohibitive difficulties, and further approximations need to be introduced.

Diagrams having one or more passive operator lines are calculated at leading order only. Such an approximation is justified by the observation that operator correlations are much weaker than the scalar ones. It would be tempting to conclude that the leading order amounts to dressing the interaction line with all possible RFHNC two-body distribution functions. However, this does not turn out to be the case. As pointed out in Section 5.2.2, describing the behaviour of correlations at intermediate and long range—driven mainly by tensor correlations and, to some extent, also by exchange correlations—requires the that nodal, or chain, diagrams be summed to infinite order.

This issue is taken care of by the so-called RFHNC/SOC approximation, which amounts to summing up the contributions of Single Operator Chains (SOC). These are chain diagrams in which any passive bond of the chain has a single operator—of the form $f^c(r_{ij})f^p(r_{ij})O^p_{ij}$, with $p \leq 6$, or $-h(r_{ij})\ell(k_F r_{ij})P_{ij}$, with $h(r_{ij}) = f^c(r_{ij})^2 - 1$—or the RFHNC-dressed version of it. Note that if a single bond of the chain is of scalar type the spin trace of the corresponding cluster term vanishes, as Pauli matrices are traceless. It follows that the SOC is the leading order, and also includes the main features of the long range behaviour of tensor and exchange correlations.

The calculation of SOC, as well as that of RFHNC chains, is based upon the convolution integral of the functions corresponding to two consecutive bonds. Unlike RFHNC chains, however, the SOC comprise operator bonds. Therefore, the basic algorithm is the convolution of two operator correlations having one common point, Eq. (5.53).

The ordering of the operators within an SOC is irrelevant, because the commutator $[O_{ik}, O_{kj}]$ is linear in $\boldsymbol{\sigma}_k$ and $\boldsymbol{\tau}_k$, and Pauli matrices are traceless. The only orderings that matter are those of the passive bonds connected to the interacting points 1 or 2, treated according to Eqs. (5.55), (5.56), (5.59), and (5.62).

A second important contribution included in the RFHNC/SOC approximation is the leading order of the vertex corrections. These are the sums of the contributions associated with sets of subdiagrams which are attached to the basic diagrammatic structure in a single point, see, e.g., diagram (b) of Fig. 5.10. In the RFHNC/SOC approximation vertex corrections are taken into account only at leading order, that is, including SOR.

The full set of RFHNC/SOC equations including SOR vertex corrections can be found in the literature [56, 118]. For the purpose of illustration, here we only discuss the equations for SOC diagrams. The contributions of nodal and composite diagrams will be denoted N^p_{xy}, X^p_{xy}, where the additional index, p, specifies the operator dependence. Elementary diagrams will be neglected altogether, according the RFHNC/0 approximation.

The generalization of Eq. (5.15) to nodal diagrams involving operators reads

$$N^r_{xy}(r_{12}) = \sum_{p,q=1}^{6} \sum_{x'y'} \text{CTr}_3\, \varrho \int_V d\mathbf{r}_3 X^p_{xx'}(r_{13}) \xi^{pqr}_{132} \zeta_{x'y'} [X^q_{y'y}(r_{23}) + N^q_{y'y}(r_{23})] \,. \tag{5.63}$$

The partial pair distribution functions, $g^p_{xy} = N^p_{xy} + X^p_{xy}$, are given by the equations

$$\begin{aligned} g^p_{dd}(r_{12}) &= h^p(r_{12}) h^c(r_{12}) \\ g^p_{de}(r_{12}) &= g^p_{ed}(r_{12}) = h^c(r_{12})[h^p(r_{12}) N^c_{de}(r_{12}) + f^c(r_{12})^2 N^p_{de}(r_{12})] \\ g^p_{ee}(r_{12}) &= h^c(r_{12})\{h^p(r_{12})[N^c_{ee}(r_{12}) + N^c_{de}(r_{12})^2] - \nu f^c(r_{12})^2 \mathcal{L}^2(r_{12}) \Delta^p \\ &\quad + N^p_{ee}(r_{12}) + 2N^c_{de}(r_{12}) N^p_{de}(r_{12})\} \,, \end{aligned} \tag{5.64}$$

where

$$\begin{aligned} h^p(r_{12}) &= 2f^p(r_{12}) f^c(r_{12}) + f^c(r_{12})^2 N^p_{dd}(r_{12}) \,, \\ h^c(r_{12}) &= \exp[N_{dd}(r_{12})] \,, \\ \mathcal{L}(r_{12}) &= N_{cc}(r_{12}) - \frac{1}{\nu}\, \ell(r_{12}) \,, \end{aligned} \tag{5.65}$$

with $\Delta^{p\leq 4} = 1$, and $\Delta^{p=5,6} = 0$. Equations (5.64) are the generalisation of the corresponding results obtained in the case of Jastrow correlations, Eqs.(5.29).

The composite functions can be readily obtained by subtracting the contribution of the nodal diagrams from the partial two-body distribution functions

$$X^p_{xy}(r_{12}) = g^p_{xy}(r_{12}) - N^p_{xy}(r_{12}) \,. \tag{5.66}$$

The total operator distribution function is given by

$$g^p(r_{12}) = g^p_{dd}(r_{12}) + 2g^p_{de}(r_{12}) + g^p_{ee}(r_{12})\,. \tag{5.67}$$

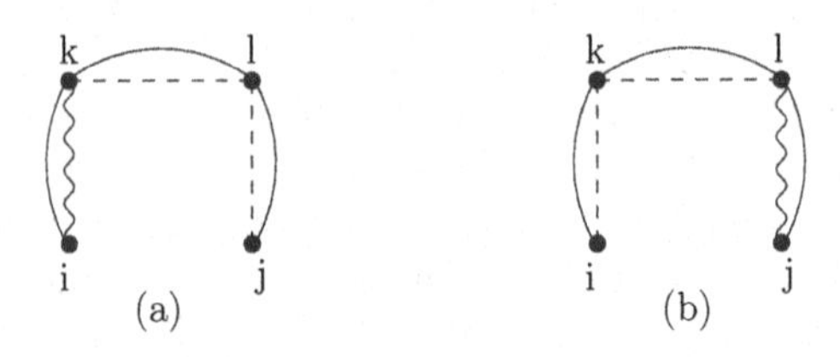

Figure 5.11 Nodal diagrams with a cyclic exchange loop.

In a generic exchange loop all links but one involve an operator dependence, and the only gap is filled by a dynamical operator completing the operator chain. Within the SOC approximation there are two distinct possibilities: the dynamical operator may be inserted either to the left (L) or to right (R) of the chain, as in the nodal diagrams (a) and (b) of Fig. 5.11, respectively. Dealing with a cyclic exchange requires the introduction of the new bond

$$X^p_{c\,L,R}(r_{12}) = h^c(r_{12})[h^p(r_{12})\mathcal{L}(r_{12}) + f^c(r_{12})^2 N^p_{c\,L,R}(r_{12})] - N^p_{c\,L,R}(r_{12})\,, \tag{5.68}$$

and the corresponding nodal functions read

$$\begin{aligned}
N^r_{cL}(r_{12}) &= \sum_{p,q=1}^{6} \mathrm{CTr}_3\, \varrho \int d\mathbf{r}_3 X^p_{c\,L}(r_{13}) \xi^{pqr}_{132} \Delta^q [X_{cc}(r_{23}) + \mathcal{L}(r_{23})] \\
N^r_{cR}(r_{12}) &= \sum_{p,q=1}^{6} \mathrm{CTr}_3\, \varrho \int \mathbf{r}_3 X_{cc}(r_{13}) \Delta^p \xi^{pqr}_{132} [X^q_{c\,R}(r_{13}) + N^q_{c\,R}(r_{13})] \\
N^r_{cc}(r_{12}) &= N^r_{cL}(r_{12}) + N^r_{cR}(r_{12})\,.
\end{aligned} \tag{5.69}$$

Finally, the partial two-body distribution functions associated with circular exchanges is given by

$$g^p_c(r_{12}) = g^c_{cc}(r_{12})\Delta^p\,. \tag{5.70}$$

The cc nodal functions enter as closed SOR in the generalised equations for X^c_{ee} and N^c_{cc}. Moreover, they contribute to the expectation value of the Hamiltonian.

5.3.3 Determination of the correlation functions

An upper bound to the ground-state energy per particle of nuclear matter, E_0/A, can be obtained by using the variational method, which amounts to minimising the expectation value of the Hamiltonian $\langle H\rangle/A$ with respect to the variational parameters included in the trial ground-state wave function. Within the cluster expansion formalism

$$\frac{\langle H\rangle}{A} = T_F + (\Delta E)_2 + \text{higher order terms}\ , \tag{5.71}$$

where T_F is the energy per particle of the non interacting Fermi gas and $(\Delta E)_2$ denotes the contribution of two-nucleon clusters

$$(\Delta E)_2 = F^\dagger v_{12} F \Big|_{2b} + \langle T \rangle \Big|_{2b} . \tag{5.72}$$

Neglecting higher order cluster contributions, the functional minimisation of $\langle H \rangle / A$ leads to a set of six Euler-Lagrange equations, to be solved with the constraints needed to enforce the proper behaviour of the correlation functions—that is, $f^c(r) \to 1$ and $f^{(p>1)}(r) \to 0$—at large distance r. These requirements are implemented through the boundary conditions

$$f^p(r \geq d^p) = \delta_{p1} \quad , \quad \left(\frac{df^p(r)}{dr} \right)_{r=d^p} = 0 \ , \tag{5.73}$$

with $p = 1, \ldots, 6$. Numerical calculations are generally carried out using only two independent parameters determining the ranges of central and tensor correlations, denoted $d_c = d^{p=1 \leq 4}$, and $d_t = d^{5,6}$, respectively.

Additional variational parameters employed in the literature also include the quenching factors α_p, meant to account for modifications of the two–body potentials induced by the presence of the nuclear medium through the replacement

$$v_{ij} \to v'_{ij} = \sum_{p=1}^{6} \alpha_p v^p(r_{ij}) O^p_{ij} \ , \tag{5.74}$$

In addition, the correlation functions are often rescaled according to

$$F_{ij} = \sum_{p=1}^{6} \beta_p f^p(r_{ij}) O^p_{ij} \ . \tag{5.75}$$

The energy expectation value $\langle H \rangle / A$, calculated within the full RFHNC/SOC approximation, is minimised with respect to variations of d_c, d_t, and the sets $\{\alpha_p\}$ and $\{\beta_p\}$, with $p = 1, \ldots 6$.

The typical shape of the correlation functions obtained from the above procedure is illustrated in Fig. 5.12. The results have been obtained using a nuclear Hamiltonian comprising the Argonne v_{14} (AV14) NN potential and the UVII model of the NNN potential.

5.3.4 Applications to the study of nuclear matter properties

Ground state energy of SNM and PNM

The RFHNC/SOC approximation has been extensively employed to carry out calculations of nuclear matter [120, 121]. Figures 5.13 and 5.14 show the results reported in the classic paper of A. Akmal, V. Pandharipande and G. Ravenhall [121]. These studies have been carried out using the AV18+UIX Hamiltonian, with and without the relativistic boost correction δv discussed in Section 4.7.1. Note that the inclusion of δv leads to a redefinition of the NNN potential, with the updated version being labelled UIX′. For comparison, the results of RFHNC/SOC calculations performed using a somewhat simplified dynamical model, based on the Urbana v_{14} NN potential, supplemented with a density-dependent potential meant to account for many-nucleon interactions [122], are also shown. This model will be referred to as U14+DDI.

The AV18 + δv + UIX′ results include a density-dependent estimate of the CBF corrections to the ground-state energy. In SNM, the highest value of this correction, adjusted to

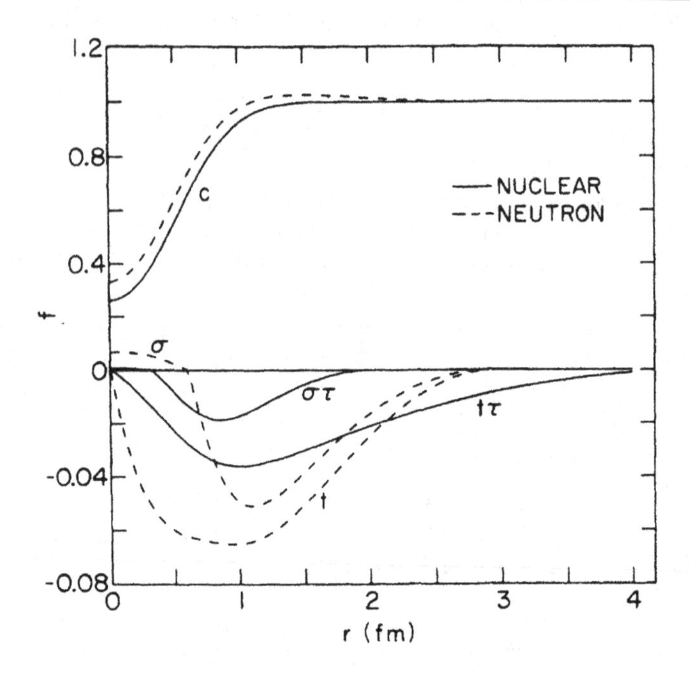

Figure 5.12 The solid and dashed line correspond to the correlations functions in SNM and PNM, respectively, at density $\varrho = 0.15$ fm^{-3}. Both sets have been obtained using a nuclear Hamiltonian including the AV14 NN potential and the and UVII NNN potential. Taken from [119].

reproduce the empirical saturation properties, turns out to be 4.5 MeV—corresponding to 30% of the interaction energy—at subnucelar density, $\varrho = 0.11$ fm^{-3}, and rapidly decreases to become negligible in the density region relevant to neutron stars. In the literature, the Hamiltonian including the boost-corrected AV18 NN potential and the UIX′ NNN potential is often referred to as the APR2 model.

A remarkable feature of the results of Figs. 5.13 and 5.14 is the size of relativistic boost corrections, which are clearly visible at $\varrho \gtrsim 2\varrho_0$ and become very large at higher densities.

Two-point Green's function

Besides the capability to systematically improve upon the variational estimates of the ground-state energy, the CBF formalism provides a consistent framework to carry out calculations of different nuclear matter properties.

The two-point Green's function and the corresponding spectral functions, defined in Section 3.3.1, have been computed for SNM at equilibrium density, using the U14+DDI Hamiltonian introduced in the previous section, and a set of correlated states $|n\rangle$, orthogonalised through a LS transformation [123, 124].

In these studies, the hole and particle contributions to the spectral function defined by

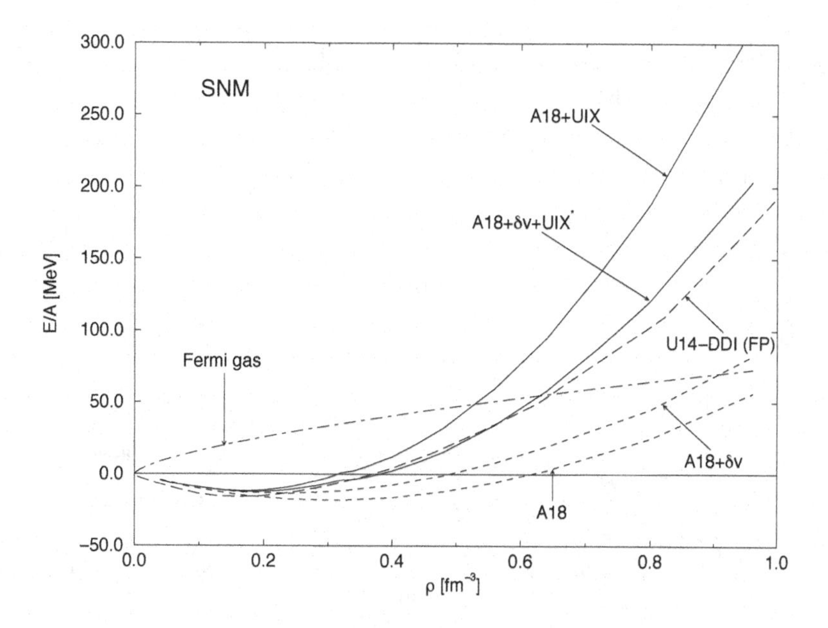

Figure 5.13 Density dependence of the ground state energy per particle of SNM, obtained within the variational RFHNC/SOC approach including the relativistic boost correction to the NN potential. The meaning of the curves is explained in the text.Taken from [121]. .

Eq. (3.54) have been written in the form

$$P_h(\mathbf{k}, E) = \sum_n |\langle n|a_{\mathbf{k}}|0\rangle|^2 \delta[E - (E_n - E_0)] \ , \tag{5.76}$$

$$P_p(\mathbf{k}, E) = \sum_h |\langle n|a^\dagger_{\mathbf{k}}|0\rangle|^2 \delta[E - (E_n - E_0)] \ , \tag{5.77}$$

where the sums in Eqs. (5.76) and (5.77) comprise orthogonal correlated states of $\mathrm{A}-1$ and $\mathrm{A}+1$ nucleons, respectively[3], the energy of which is denoted E_n. Numerical calculations of $P_h(\mathbf{k}, E)$ have been performed including one-hole and two-hole–one–particle intermediate states, while one-particle and two-particle–one-hole states have been taken into account for $P_p(\mathbf{k}, E)$.

Perturbative CBF corrections have been included to account for the coupling between intermediate states, as well as to improve the description of the variational ground state, allowing for the admixture of two-particle–two-hole states.

A three-dimensional representation of the CBF hole spectral function of SNM at equilibrium density is displayed in Fig. 5.15. The peaks corresponding to transitions to bound one-hole states of the $(\mathrm{A}-1)$-nucleon system, broadened by the mixing with two-hole–one-particle states, are complemented by a sizeable smooth component, arising from ground-state correlations. This background extends to values of $|\mathbf{k}|$ and E way beyond the typical values predicted by the independent particle model.

[3]A comparison with Eqs. (3.46)-(3.54) shows that in Eq. (5.76) $E = -\omega$ is the positive removal energy of a nucleon of momentum $\mathbf{k}$. This definition is more convenient in view of the use of the hole spectral function for the analysis of nucleon knockout experiments.

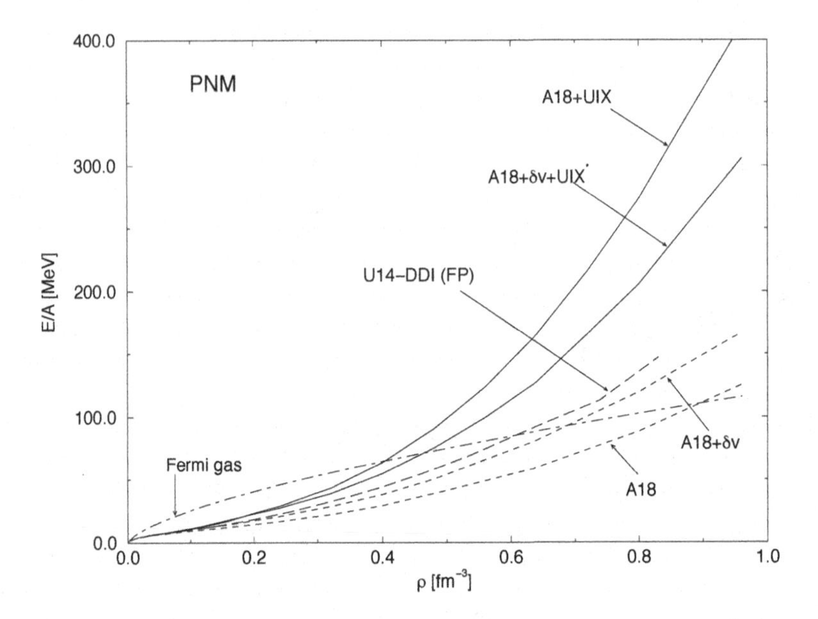

Figure 5.14 . Density dependence of the ground state energy per particle of PNM, obtained within the variational RFHNC/SOC approach including the relativistic boost correction to the NN potential. The meaning of the curves is explained in the text Taken from [121]. .

The energy dependence of the hole and particle spectral functions at momentum $|\mathbf{k}| = 0.787, 1.226, 1.797$, and 2.120 fm^{-1} is illustrated in Fig. 5.16. It is apparent that the peaks of the hole spectral function only occur at $|\mathbf{k}| < k_F = 1.33$ fm^{-1}, and their width decreases with increasing $|\mathbf{k}|$. This pattern is mirrored by the particle spectral function at $|\mathbf{k}| > k_F$. The contribution arising from ground-state correlations, on the other hand, is continuous across the Fermi surface. Note that at 0-th order of CBF, that is, at variational level, the peaks shrinks to a collection of of δ-functions located at $E = -e_{\mathbf{h}}$—with $e_{\mathbf{h}}$ being the energy of a nucleon carrying momentum $\mathbf{h}$ with $|\mathbf{h}| < k_F$—and the continuum contribution, which would disappear altogether in any independent particle models, is strongly reduced. For comparison, the 0-th order CBF spectral functions are shown by the dashed lines.

5.4 CBF EFFECTIVE INTERACTION

As pointed out in Section 4.2, the two approaches developed to overcome the problem posed by the non perturbative nature of the repulsive core of the NN potential are based either on the replacement of the bare NN potential with an effective interaction—obtained by summing up the contributions of ladder diagrams to all orders of perturbation theory—or on the replacement of the Fermi gas basis with a basis of correlated states, embodying non-perturbative interaction effects. In the early 2000s it has been suggested that these two approaches may, in fact, be combined.

In CBF perturbation theory, one has to evaluate matrix elements of the bare nuclear Hamiltonian, the effects of correlations being taken into account by the transformation of the basis states describing the non interacting system. However, the same result can in

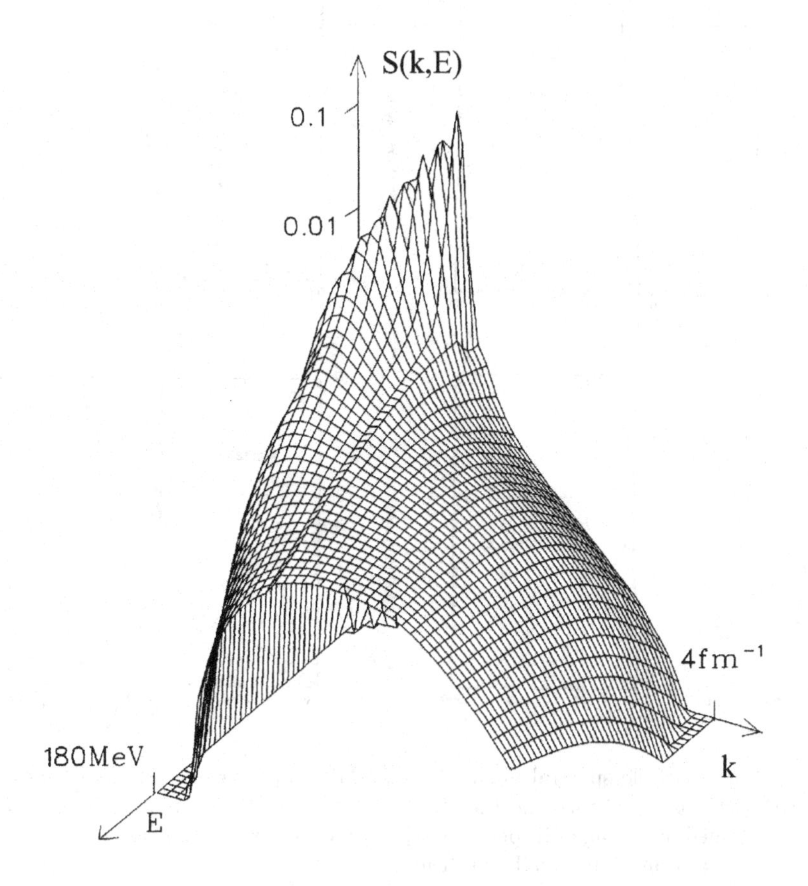

Figure 5.15 Hole spectral function of SNM at equilibrium density, corresponding to $k_F = 1.33$ fm^{-1}, obtained using orthogonal CBF perturbation theory and the U14+DDI nuclear Hamiltonian. Taken from [125].

principle be obtained transforming the Hamiltonian, and using the Fermi gas basis. This procedure leads to the appearance of an effective Hamiltonian suitable for use in standard perturbation theory, thus avoiding the non trivial difficulties arising from the use of a non-orthogonal basis [112].

The CBF effective interaction is defined through the matrix element of the bare Hamiltonian in the correlated ground state, according to

$$\frac{1}{A}\langle\Psi_T|H|\Psi_T\rangle = T_F + \frac{!}{A}\langle\Phi_0|\sum_{i<j} v_{ij}^{\text{eff}}|\Phi_0\rangle\ , \tag{5.78}$$

where $|\Psi_T\rangle$ and $|\Phi_0\rangle$ denote the correlated trial ground state and the Fermi gas ground state, respectively, T_F is the energy per particle of the non interacting Fermi gas, and the effective potential is written in terms of the same spin-isospin operators appearing in Eq. (2.12) as

$$v_{ij}^{\text{eff}} = \sum_p v^{\text{eff},p}(r_{ij})O_{ij}^p\ . \tag{5.79}$$

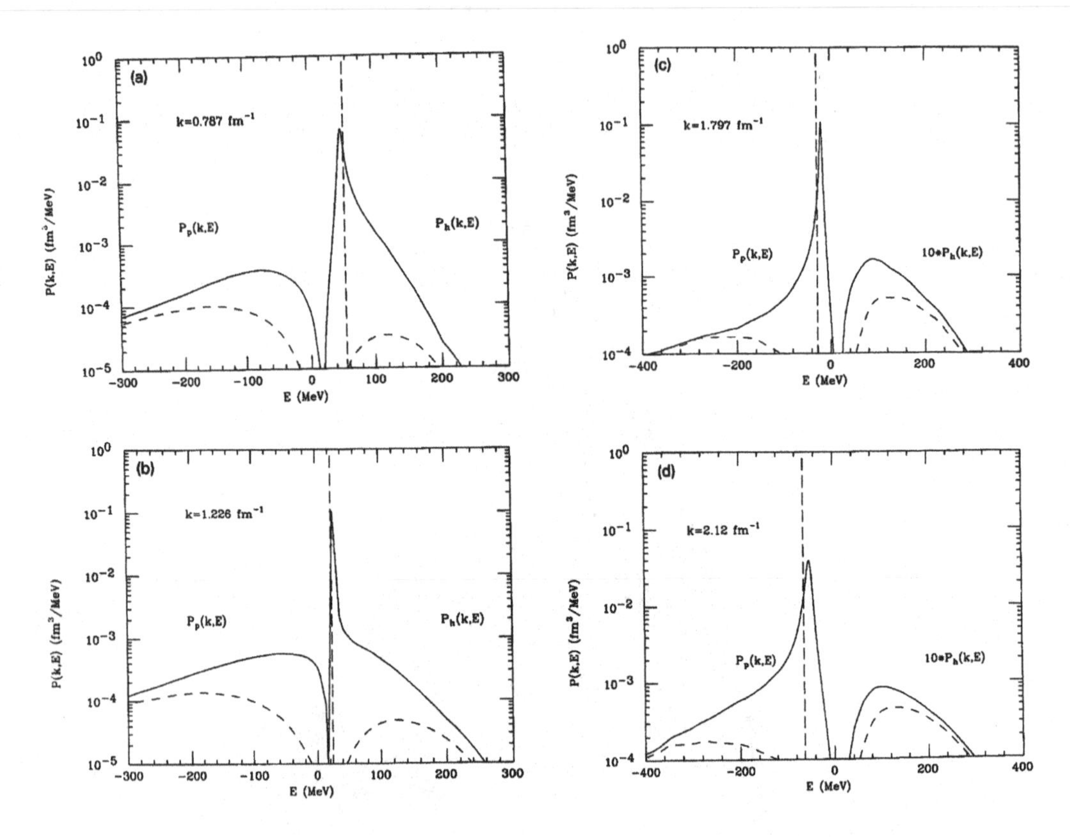

Figure 5.16 Hole and particle spectral functions of SNM at equilibrium density, computed using orthogonal CBF perturbation theory and the U14+DDI nuclear Hamiltonian. The full lines represent the results obtained including CBF perturbative corrections, while the dashed lines correspond to the 0-th, that is variational, results. Taken from [124]. .

From the above equations, it is apparent that $v^{\rm eff}_{ij}$ embodies the density-dependent effects of correlations. As a consequence, it is well behaved at short distances, and can in principle be used to carry out perturbative calculations of any properties of nuclear matter.

The basic requirement for the derivation of the effective interaction is that the left-hand side of Eq. (5.78) be accurately calculable, using either the RFHNC/SOC summation scheme or Monte Carlo techniques.

In early works [126, 127], the effective interaction was obtained performing a cluster expansion of the left-hand side of Eq. (5.78) and keeping the two-body cluster contribution only. While leading to a very simple and transparent expression for $v^{\rm eff}_{ij}$, however, this scheme is seriously limited by its inability to take into account the NNN potential V_{ijk}.

A significant improvement has been achieved taking into account three-nucleon cluster contributions [128, 129] This procedure allows to describe the effects of three-nucleon interactions using microscopic potentials, such as the UIX model.

Note that the correlation functions $f^p(r_{ij})$ entering the definition of $v^{\rm eff}_{ij}$ are not the same as those obtained from the minimization of the variational energy of Eq. (4.40). They are adjusted so that the ground state energy computed at first order in $v^{\rm eff}_{ij}$—that is, in the Hartree-Fock approximation—reproduces the value of E_V resulting from the full RFHNC/SOC calculation. Note that the effective interaction simultaneously describes the density dependence of the energy per nucleon of both SNM and PNM. This feature is very important, because it allows to evaluate the properties of nuclear matter at fixed density and large neutron excess, which is believed to make up a large region of the neutron star interior.

The radial dependence of the spherically-symmetric component of the effective potential describing the interaction of two nucleons coupled with total spin and isospin $S = 1$ and $T = 0$ is illustrated in Fig. 5.17. The calculation has been carried out at SNM equilibrium density, using a nuclear Hamiltonian including the AV6P and UIX potentials. Comparison with the bare V6P potential clearly shows that correlations significantly affect both the short- and intermediate-range behaviour.

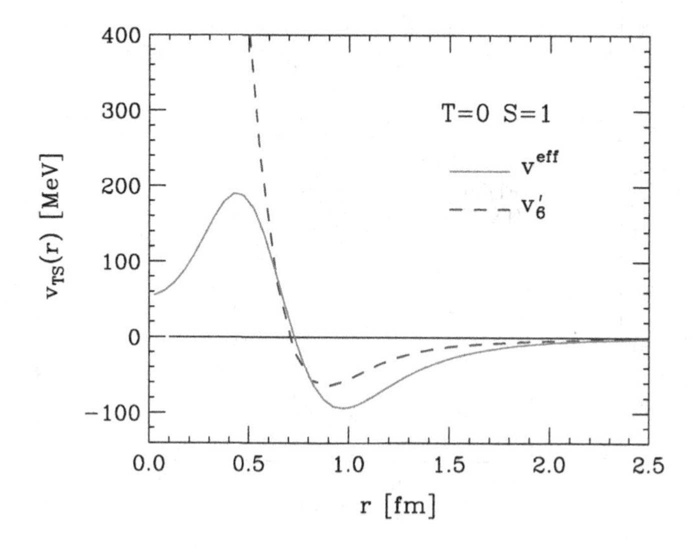

Figure 5.17 Radial dependence of the spherically-symmetric component of the bare AV6P potential (dashed line) and the CBF effective interaction (solid line) in the spin-isospin channel corresponding to $S = 1$ and $T = 0$. The effective interaction has been computed setting $\varrho = \varrho_0 = 0.16$ fm^{-3}. Taken from [60]. .

The solid lines of Figs. 5.18 illustrate the density dependence of the energy per nucleon of PNM and SNM, obtained from the CBF effective interaction. The shaded regions show the RFHNC/SOC results obtained from the bare Hamiltonian, with the associated theoretical uncertainty arising from the treatment of the kinetic energy, discussed in Chapter 4. For comparison, the results of a calculation carried out using the AFDMC method—see Section 4.6.2—are also displayed. It clearly appears that the RFHNC/SOC variational estimates, exploited as baseline for the determination of the CBF effective interaction, provide very accurate upper bounds to the ground state energy of PNM over the whole density range. Note that the simplified AV6P+UIX Hamiltonian yields the correct equilibrium density of SNM, $\varrho_0 \approx 0.16$ fm^{-3}, although the corresponding binding energy, ~ 11 MeV, is below the empirical value of 16 MeV. However, it must be kept in mind that, because the kinetic and interaction energies largely cancel one another, a ~ 5 MeV discrepancy in the ground-state energy translates into a $\sim 15\%$ underestimate of the interaction energy. This is consistent with the results of variational calculations of SNM performed with the full AV18+UIX Hamiltonian, yielding $E_0/A = -11.85\,\text{MeV}$.

The compressibility module, symmetry energy and pressure obtained from the CBF

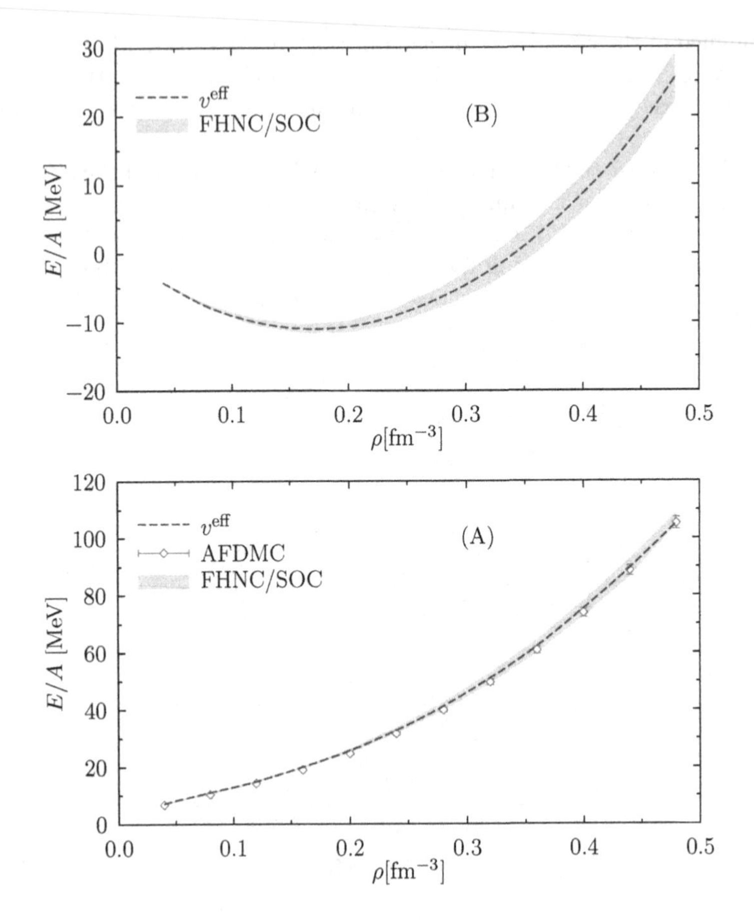

Figure 5.18 Density dependence of the energy per nucleon of PNM (A) and SNM (B). The dashed lines show the results obtained using the CBF effective interaction.The variational RFHNC/SOC results are represented by the shaded regions, illustrating the uncertainty associated with the treatment of the kinetic energy, while the open circles of panel (A) correspond to the PNM results obtained using the AFDMC technique. Taken from [60]. .

effective interactions turn out to be compatible with the empirical information discussed in Chapter 3 [60].

As pointed out above, using the CBF effective interaction the calculation of the ground-state energy of nuclear matter with arbitrary proton fraction does not involve any significant additional difficulties. The results corresponding to fixed baryon density ϱ and proton density $\varrho_p = x_p\varrho$, with x_p in the range $0 \leq x_p \leq 0.5$ are displayed in Fig. 5.19.

The derivation of the effective interaction based on the CBF formalism is conceptually similar to the renormalisation group evolution of the bare NN potential, in that both methods are designed to take into account the screening of nuclear interactions in the nuclear medium. While in the renormalisation group approach the evolution is driven by a momentum cutoff, in the CBF approach screening is described in coordinate space, and the evolution is driven by the matter density.

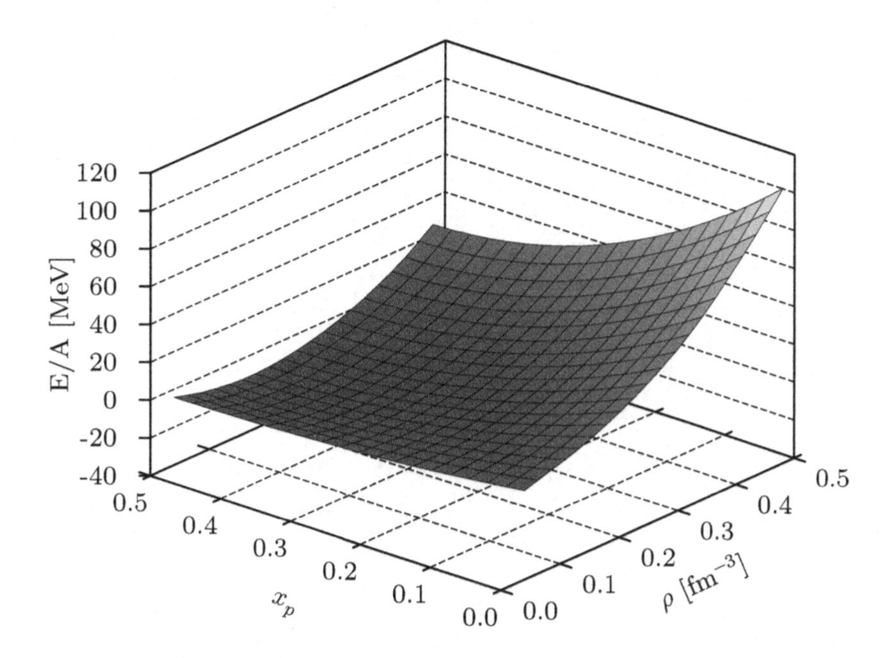

Figure 5.19 Energy per nucleon of nuclear matter, computed as a function of baryon density and proton fraction using the CBF effective interaction. Taken from [60]. .

CHAPTER 6

NEUTRON STARS

The appearance of a neutron star is one of the possible outcomes of stellar evolution. In this chapter we briefly review the processes leading to the formation of a neutron star and the structure of matter in its interior, and discuss the main properties of the region in which the description based on nuclear matter theory is expected to be applicable. The derivation of the equations of hydrostatic equilibrium in general relativity, providing a link between the microscopic dynamics determining the EOS of nuclear matter and the neutron star mass and radius, is also outlined.

6.1 NEUTRON STAR FORMATION

The formation of a star is triggered by the contraction of a self-gravitating hydrogen cloud. As the density increases, so does the opacity of the cloud, and the energy released cannot be efficiently radiated away. As a consequence, the temperature grows, until it reaches the value $T \sim 6 \times 10^7$ K, needed to ignite the cycle of nuclear reactions turning hydrogen into helium

$$
\begin{aligned}
p + p &\rightarrow {}^2\mathrm{H} + e^+ + \nu + 0.4 \ \mathrm{MeV} , \\
e^+ + e^- &\rightarrow \gamma + 1.0 \ \mathrm{MeV} , \\
{}^2\mathrm{H} + p &\rightarrow {}^3\mathrm{He} + \gamma + 5.5 \ \mathrm{MeV} , \\
{}^3\mathrm{He} + {}^3\mathrm{He} &\rightarrow {}^4\mathrm{He} + 2p + 26.7 \ \mathrm{MeV} .
\end{aligned}
$$

The above reactions are all exothermic, and energy is released in the form of kinetic energy of the produced particles. Equilibrium is achieved as soon as gravitational attraction is balanced by matter pressure.

When the nuclear fuel is exhausted, the core stops producing heat, the internal pressure can no longer be sustained, and the gravitational pull resumes. If the mass of the helium core is large enough, its contraction, associated with a further increase of the temperature, can then lead to the ignition of a new cycle of fusion reactions, resulting in the appearance of heavier nuclei, such as carbon and oxygen. Depending on the mass of the initial configuration, M_0, this process can take place several times, the end point being the formation of a core made of the most stable nuclear species, nickel and iron, see Fig. 1.1, at density $\sim 10^{14}$ g/cm^3. The stages of nucleosynthesis for a star with $M_0 = 25\ M_\odot$ are summarized in Table 6.1.

The outcome of stellar evolution is largely driven by the value of M_0. If the star is

Nuclear fuel	Main products	Temperature [K]	Density [g/cm^3]	Duration [yrs]
H	He	3.81×10^7	3.81	6.7×10^6
He	C, O	1.96×10^8	762	8.39×10^5
C	O, Ne, Mg	8.41×10^8	1.29×10^5	522
Ne	O, Mg, Si	1.57×10^9	3.95×10^6	0.891
O	Si, S	2.09×10^9	3.60×10^6	0.402
Si	Fe	3.65×10^9	3.01×10^7	0.002

Table 6.1 Stages of nucleosynthesis for a star with initial mass $M_0 = 25\ M_\odot$. Data from [130].

not very massive, typically for $M < 10\ M_\odot$, and the gravitational contraction of the core does not lead to reaching the temperatures needed to ignite the burning of heavy nuclei, it will eventually turn into a white dwarf, consisting mainly of fully ionized helium, carbon and oxygen. The measured values of mass and radius of white dwarfs are $\sim 1\ M_\odot$ and ~ 5000 km, respectively, and the corresponding average density is $\sim\ 10^6$ g/cm^3.

The pressure required to ensure the stability of white dwarfs against gravitational collapse is provided by the strongly degenerate electron gas. Based on this conjecture, in 1931 S. Chandrasekhar was able to predict the existence of a maximum mass, $M_C \approx 1.44\ M_\odot$, for white dwarfs [131]. Beyond this limiting value, referred to as Chandrasekhar mass, gravitational attraction prevails on the pressure gradient, and the star becomes unstable against gravitational collapse. It is truly remarkable that the stability of white dwarfs turns out to be the macroscopic manifestation of a purely quantum mechanical effect, the degeneracy pressure of the Fermi gas arising from Pauli's exclusion principle.

In stars having initial mass larger than $\sim 10\ M_\odot$, the temperatures can be high enough to bring nucleosynthesis to the final stage, thus allowing the development of an iron core, see Table 6.1. If the mass of the core exceeds the Chandrasekhar mass, the pressure of the electron gas cannot balance the gravitational contraction, and the star evolves towards the formation of a neutron star or a black hole.

The formation of the core in massive stars is characterised by the emission of neutrinos, produced in the reaction

$$^{56}Ni \ \rightarrow\ ^{56}Fe + 2e^+ + 2\nu_e \ . \tag{6.1}$$

Neutrinos do not have appreciable interactions with the surrounding matter. They leave the core region carrying away energy, thus contributing to the collapse of the core. Additional processes leading to a decrease of the pressure are electron capture

$$e^- + p \ \rightarrow\ n + \nu_e \ , \tag{6.2}$$

whose main effect is the disappearance of relativistic electrons, carrying large kinetic energies, and iron photo-disintegration

$$\gamma + {}^{56}\mathrm{Fe} \ \rightarrow\ 13\ {}^4\mathrm{He} + 4n \ , \tag{6.3}$$

which is an endothermic reaction.

Owing to the combined effect of the above processes, when its mass exceeds the Chandrasekhar limit the core collapses, reaching densities as high as $\sim 10^{14}$ g/cm^3—a value comparable with the central density of atomic nuclei—within fractions of a second.

At this stage, the core behaves as a giant nucleus, consisting mostly of neutrons, and reacts elastically to further compression, producing a strong shock wave which ejects a significant fraction of matter in the outer layers of the star. Nucleosynthesis of elements heavier than iron is believed to take place during this phase, referred to as supernova explosion.

The supernova is a star whose luminosity rapidly grows to reach a maximum value exceeding the luminosity of the sun by a factor $\sim 10^9$, and then decreases by a factor $\sim 10^2$ within few months. For $M_0 < 30\ M_\odot$ the final result of the explosion is the formation of a *nebula*, the centre of which is occupied by the remnant of the core, that is, a neutron star. The collapse of stars with larger initial mass is believed to lead to the formation of a black hole.

6.2 NEUTRON STAR STRUCTURE

6.2.1 Crust region

The neutron star structure, briefly sketched in Chapter 1, features a succession of layers, consisting of different forms of matter, spanning a density range of more than seven orders of magnitude, see Fig. 1.4. For any baryon density ϱ, matter composition is determined by the condition of minimum energy, with the additional constraints of charge neutrality and β-stability.

The outer crust comprises a lattice of heavy nuclei, surrounded by a strongly degenerate electron gas. Moving from the surface towards the interior of the star the density increases, and so does the electron kinetic energy. As a consequence, the electron capture reaction

$$p + e \rightarrow n + \nu_e \ , \tag{6.4}$$

leading to the appearance of more and more neutron-rich nuclei, becomes very efficient. Table 6.2 shows the sequence of nuclides corresponding to the ground state of matter at different densities. At $\varrho \sim 4 \times 10^{11}$ g/cm^3 the ground state corresponds to a Coulomb lattice of ^{118}Kr nuclei, having proton to neutron ratio $(Z/N) \sim 0.31$ and a slightly negative neutron Fermi energy.

At larger densities a new regime sets in, since the neutrons produced by electron captures can no longer be accommodated in bound, negative energy, states, and begin to drip out of the nuclei. In this region, called inner crust, the ground state of matter is a mixture of two phases: neutron-rich nuclei at density $\varrho_{\rm nuc}$, or phase I, and a neutron gas at density $\varrho_{\rm NG}$, or phase II.

The details of the structure of the inner crust are determined by the densities ϱ, $\varrho_{\rm nuc}$ and $\varrho_{\rm NG}$, the proton to neutron ratio of matter in phase I, and the interplay between the nuclear surface energy, which is reduced by the presence of the neutron gas, and the Coulomb interaction between nuclei.

Theoretical studies suggest that at densities $4.3\times10^{10} \lesssim \varrho \lesssim .75\times10^{14}$ g/cm^3 matter in phase I is arranged in spheres surrounded by the electron and neutron gases, whereas at $.75\times10^{14} \lesssim \varrho \lesssim 1.2\times10^{14}$ g/cm^3 the energetically favoured configurations consist of rods of matter in phase I or alternating layers of matter in phases I and II. These forms of matter, schematically represented in Fig. 6.1, are referred to as nuclear pasta [132, 133].

Finally, as the density reaches values slightly below ϱ_0—marking the boundary between the inner crust and the neutron star core—nuclear pasta dissolves into a uniform fluid of

Nuclide	Z	A − Z	Z/A	Δ [MeV]	ϱ [g/cm^3]
^{56}Fe	26	30	.4643	.1616	8.1×10^6
^{62}Ni	28	34	.4516	.1738	2.7×10^8
^{64}Ni	28	36	.4375	.2091	1.2×10^9
^{84}Se	34	50	.4048	.3494	8.2×10^9
^{82}Ge	32	50	.3902	.4515	2.1×10^{10}
^{84}Zn	30	54	.3750	.6232	4.8×10^{10}
^{78}Ni	28	50	.3590	.8011	1.6×10^{11}
^{76}Fe	26	50	.3421	1.1135	1.8×10^{11}
^{124}Mo	42	82	3387	1.2569	1.9×10^{11}
^{122}Zr	40	82	.3279	1.4581	2.7×10^{11}
^{120}Sr	38	82	.3166	1.6909	3.7×10^{11}
^{118}Kr	36	82	.3051	1.9579	4.3×10^{11}

Table 6.2 Sequence of nuclei corresponding to the ground state of matter at density ϱ. The quantity Δ is defined by the equation M(Z,A)/A = (930 + Δ) MeV, with M(Z,A) being the nuclear mass. Data from Ref. [134].

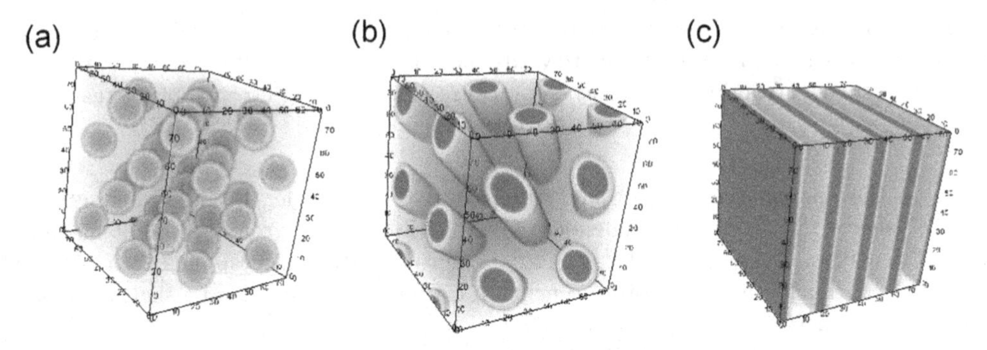

Figure 6.1 Schematic representation of the nuclear pasta phase in the neutron star inner crust. Panels (a), (b) and (c) represent droplets of matter in phase I, cylindrical rods of matter in phase I, and alternate slabs of matter in phases I and II, respectively [135].

neutrons, protons, electrons and muons, which is believed to be the ground state of matter up to densities as high as two or three times the equilibrium density of SNM. Note that the production of muons in weak interaction processes becomes energetically allowed at high density, as soon as the electron kinetic energy exceeds the electron-muon mass difference $\Delta m = m_\mu - m_e \approx 105$ MeV.

6.2.2 Core region

Nuclear matter theory provides a consistent framework to describe the region of the neutron star core in which nucleons are the relevant degrees of freedom and the effects of nuclear structure are expected to be negligible. This region accounts for $\sim 99\%$ of the neutron star mass, and plays a dominant role in determining a number of neutron star properties.

At any given nucleon density ϱ, the ground state of matter in the neutron star core is determined by minimisation of the function

$$F(\varrho, \varrho_p, \varrho_n, \varrho_e, \varrho_\mu) = \epsilon(\varrho, \varrho_p, \varrho_n, \varrho_e, \varrho_\mu) \tag{6.5}$$

$$+ \lambda_B(\varrho - \varrho_p - \varrho_n) + \lambda_Q(\varrho_p - \varrho_e - \varrho_\mu) , \tag{6.6}$$

where ϱ_α, with $\alpha = n, \ p, \ e$ or μ, denotes the density of particles of species α, $\epsilon = E_0/V$ is the energy density, and λ_B and λ_Q are Lagrange multipliers, whose values are fixed by the requirements of baryon number conservation and charge neutrality.

By solving the equations

$$\frac{\partial F}{\partial \varrho_\alpha} = 0 \quad , \quad \frac{\partial F}{\partial \lambda_B} = \frac{\partial F}{\partial \lambda_Q} = 0, \tag{6.7}$$

and substituting the chemical potentials, defined as

$$\mu_\alpha = \left(\frac{\partial \epsilon}{\partial \varrho_\alpha} \right)_{V, \varrho_\beta \neq \varrho_\alpha} , \tag{6.8}$$

one finds the relations

$$\mu_n = \mu_p + \mu_e \quad , \quad \mu_\mu = \mu_e , \tag{6.9}$$

expressing chemical equilibrium with respect to the reactions

$$n \to p + \ell + \bar{\nu}_\ell \quad , \quad p + \ell \to n + \nu_\ell , \tag{6.10}$$

with $\ell = e$ or μ, as well as the conservation laws, implying

$$\varrho = \varrho_p + \varrho_n \quad , \quad \varrho_p = \varrho_e + \varrho_\mu . \tag{6.11}$$

Note that the above results, allowing to obtain the densities of matter constituents, ϱ_α, for any fixed value of ϱ, have been derived assuming that neutron stars are transparent to neutrinos. This assumption is justified by the results of theoretical studies of the neutrino mean free path in neutron star matter, defined as

$$\lambda = \frac{1}{\varrho \sigma(E_\nu)} , \tag{6.12}$$

where ϱ and $\sigma(E_\nu)$ denote the matter density and the total neutrino cross section at energy E_ν, respectively. For thermal neutrinos, that is, for $E_\nu \sim T$, λ turns out to largely exceed the typical neutron star radius as soon as the temperature drops below 10^{10} K. As a consequence, in this regime neutrinos can be treated as non degenerate particles, having vanishing density and chemical potential.

6.3 EQUATION OF STATE OF NEUTRON STAR MATTER

The determination of the EOS of charge-neutral β-stable matter involves the calculation of the ground-state energy at nucleon density ϱ and proton and neutron densities obtained from the solution of the chemical equilibrium equations (6.9) and (6.11), conveniently parametrised in terms of the proton fraction x according to $\varrho_p = x\varrho$ and $\varrho_n = (1-x)\varrho$.

Although the energy $E_0(\varrho, x)$ can, at least in principle, be computed within the framework of the theoretical approaches described in Chapters 4 and 5.1, the discussion of the nuclear matter symmetry energy of Section 3.2.2 suggests that accurate estimates can be obtained from the energies of PNM and SNM, corresponding to $x = 0$ and $x = 1/2$, respectively.

For any values of ϱ and x, the ground-state energy per nucleon can be rewritten in the form

$$\frac{1}{A} E_0(\varrho, x) = K(\varrho, x) + V(\varrho, x) \ , \tag{6.13}$$

where $K(\varrho, x)$ is the energy per particle of a two-component Fermi gas of degeneracy $\nu = 2$

$$K(\varrho, x) = x K_p(\varrho, x) + (1-x) K_n(\varrho, 1-x) \ , \tag{6.14}$$

with the proton and neutron contributions being given by

$$K_p(\varrho, x) = \frac{3}{10m}(3\pi^2 x\varrho)^{2/3} \quad , \quad K_n(\varrho, 1-x) = \frac{3}{10m}[3\pi^2(1-x)\varrho]^{2/3} \ . \tag{6.15}$$

The x dependence of the interaction energy $V(\varrho, x)$ can be described performing an expansion in powers of $\delta^2 = (1-2x)^2$, reminiscent of Eq. (3.34). The resulting expression is

$$V(\varrho, x) = V_0(\varrho) + V_2(\varrho)(1-2x)^2 \ , \tag{6.16}$$

where

$$V_0(\varrho) = V(\varrho, 1/2) \quad , \quad V_2(\varrho) = V(\varrho, 0) - V(\varrho, 1/2) \ . \tag{6.17}$$

For a given density ϱ, the above equations allow to obtain the interaction energy of nuclear matter for any x from the values corresponding to $x = 0$ and $1/2$, which can be obtained from the ground-state energy of PNM and SNM, respectively. Note that subtraction of the Fermi gas energy is required, since $K(\varrho, x)$ is not quadratically dependent on δ.

The proton and neutron chemical potentials appearing in Eq.(6.9) can be readily obtained from the thermodynamic definition, Eq.(6.8). Using Eqs.(6.14)-(6.17) to obtain the energy density, ϵ, one finds

$$\begin{aligned} \mu_p(\varrho, x) = & \frac{1}{2m}(3\pi^2 xn)^{2/3} + V_0(\varrho) + \varrho V_0'(\varrho) \\ & + \varrho(1-2x)^2 V_2'(\varrho) + (-4x^2 + 8x - 3) V_2(\varrho) \ , \end{aligned} \tag{6.18}$$

and

$$\begin{aligned} \mu_n(\varrho, x) = & \frac{1}{2m}[3\pi^2(1-x)\varrho]^{2/3} + V_0(\varrho) + \varrho V_0'(\varrho) \\ & + \varrho(1-2x)^2 V_2'(\varrho) + (1-4x^2) V_2(\varrho) \ . \end{aligned} \tag{6.19}$$

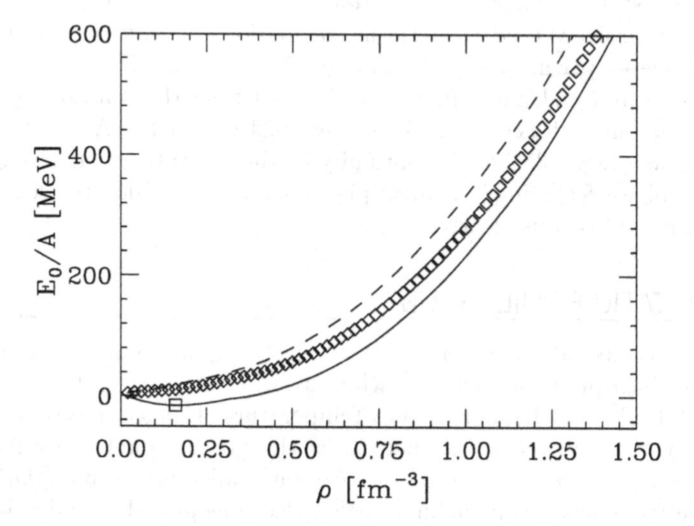

Figure 6.2 The diamonds show the density dependence of the energy per nucleon of charge-neutral β-stable nuclear matter obtained from the APR2 model of Akmal *et al.* [121]. For comparison, the corresponding SNM and PNM results are displayed by the solid and dashed lines, respectively. The square corresponds to the empirical equilibrium point of SNM.

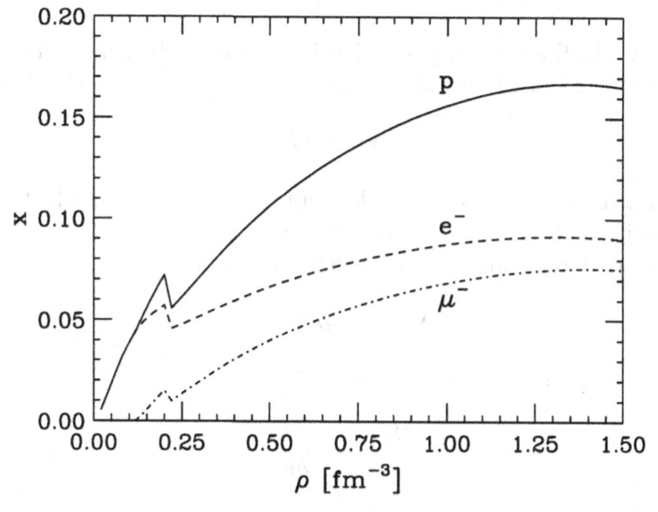

Figure 6.3 Proton, electron and muon fractions of charge-neutral β-stable matter, obtained from the APR2 model of Akmal *et al.* [121].

The charged leptons are treated as non interacting particles of mass m_ℓ, with $\ell = e, \mu$, whose chemical potential coincides with the Fermi energy

$$\mu_\ell = \sqrt{(3\pi^2 \varrho_\ell)^{2/3} + m_\ell^2} \; . \tag{6.20}$$

As an example, Fig. 6.2 illustrates the density dependence of the energy per nucleon of charge-neutral β-stable matter obtained from the APR2 model described in Section 5.3, which has been extensively used to perform calculations of neutron star properties. For comparison, the energy per nucleon of SNM and PNM are also shown.

The proton, electron and muon fractions obtained from the solution of Eq.(6.9) using the chemical potentials of Eqs.(6.18), (6.19) and (6.20)[1] and the APR2 energies of SNM and PNM are shown Fig. 6.3. The discontinuity of the derivatives, signalling the onset of the high-density phase featuring a neutral-pion condensate at density slightly above the equilibrium density of SNM, is clearly visible.

6.4 HYDROSTATIC EQUILIBRIUM

The derivation of the equations describing hydrostatic equilibrium of self-gravitating systems can be best illustrated considering a white dwarf of mass M and radius R, consisting of a plasma of fully ionised helium at zero temperature. In such a system, surface gravity—measured by the ratio GM/R, where G is the gravitational constant—is small, of order $\sim 10^{-4}$. As a consequence, its structure can be studied neglecting relativistic effects.

Owing to the large mass of the helium nuclei, electrons provide the dominant contribution to the pressure, P, that can be obtained from the EOS using Eq. (3.14).

Assuming that the system behaves as a perfect fluid in thermodynamic equilibrium, subject to gravity only, hydrostatic equilibrium is described by Euler's equation, that can be written in the form

$$\frac{\partial \boldsymbol{v}}{\partial t} + (\boldsymbol{v} \cdot \boldsymbol{\nabla})\boldsymbol{v} = -\frac{1}{\varrho}\boldsymbol{\nabla} P - \boldsymbol{\nabla}\phi \ . \tag{6.21}$$

Here, $\boldsymbol{v}$ and ϱ denote the fluid velocity and density, respectively, while ϕ is the gravitational potential, solution of the Poisson equation

$$\boldsymbol{\nabla}^2\phi = 4\pi G\varrho \ . \tag{6.22}$$

The above equation provides an accurate description of systems in which dissipative processes due to viscosity and heat transfer can be neglected.

For a fluid at rest, that is, for $\boldsymbol{v} = 0$, Eq.(6.21) reduces to

$$\boldsymbol{\nabla} P = -\varrho \, \boldsymbol{\nabla}\phi \ , \tag{6.23}$$

and exploiting spherical symmetry one finds

$$\frac{dP}{dr} = -\varrho\frac{d\phi}{dr} \ , \tag{6.24}$$

and

$$\frac{1}{r^2}\frac{d}{dr}\left(r^2\frac{d\phi}{dr}\right) = 4\pi G\varrho \ . \tag{6.25}$$

Finally, substitution of Eq. (6.24) into Eq. (6.25) yields

$$\frac{1}{r^2}\frac{d}{dr}\left(\frac{r^2}{\varrho}\frac{dP}{dr}\right) = -4\pi G\varrho \ , \tag{6.26}$$

[1] Note that the proton and neutron chemical potentials should also include a mass term. However, neglecting the small difference between proton and neutron mass these terms do not affect chemical equilibrium.

implying

$$\frac{dP}{dr} = -\varrho(r)\,\frac{GM(r)}{r^2}\,, \tag{6.27}$$

with $M(r)$ given by

$$M(r) = 4\pi \int_0^r \varrho(r')r'^2 dr'\,. \tag{6.28}$$

The above result shows that at equilibrium the gravitational attraction acting on a volume element at distance r from the centre of the star is is balanced by the force produced by the pressure gradient.

Knowing the EOS in the form $P = P(\varrho)$, Eq.(6.27) can be integrated numerically for any values of the central density ϱ_c to obtain the radius of the star, R, determined by the condition $\varrho(R) = 0$. The mass $M(R)$ can then be obtained from Eq.(6.28) setting the upper integration limit to $r = R$.

6.4.1 The equations of Tolman, Oppenheimer and Volkoff

The Newtonian treatment of equilibrium employed in the case of white dwarfs is only applicable to stars whose density does not produce a significant curvature of space-time. In these conditions, the space-time metric is simply given by

$$ds^2 = \eta_{\mu\nu} dx^\mu dx^\nu\,, \tag{6.29}$$

with $\eta^{\mu\nu} = \text{diag}(-1, 1, 1, 1)$. In Einstein's theory of general relativity, Eq.(6.29) is replaced by

$$ds^2 = g_{\mu\nu} dx^\mu dx^\nu\,, \tag{6.30}$$

where the metric tensor $g_{\mu\nu}$ is a function of space-time coordinates.

The effects of space-time distortion are negligible when the surface gravitational potential satisfies the condition $GM/R \ll 1$. This requirement is met by white dwarfs, having $GM/R \sim 10^{-4}$, but not by neutron stars, whose larger density leads to much higher gravitational potentials, typically $\sim 10^{-1}$.

Relativistic corrections to the hydrostatic equilibrium equations (6.27) and (6.28) are derived from Einstein's field equations

$$G_{\mu\nu} = 8\pi G T_{\mu\nu}\,, \tag{6.31}$$

where $T_{\mu\nu}$ is the energy-momentum tensor, and Einstein's tensor $G_{\mu\nu}$ is defined in terms of the metric tensor $g_{\mu\nu}$ describing space-time geometry. Equations (6.31) establish the relation between the distribution of matter, described by $T_{\mu\nu}$, and space-time curvature, described by $g_{\mu\nu}$.

Let us now consider a star consisting of a static and spherically simmetric distribution of matter in chemical, hydrostatic and thermodynamic equilibrium. The metric of the corresponding gravitational field can be written in the form

$$ds^2 = g_{\mu\nu} dx^\mu dx^\nu = e^{2\nu(r)} dt^2 - e^{2\lambda(r)} dr^2 - r^2(d\theta^2 + \sin^2\theta d\varphi^2)\,, \tag{6.32}$$

with $x^0 = t$, $x^1 = r$, $x^2 = \theta$, and $x^3 = \varphi$, implying $g^{\mu\nu} = \text{diag}(e^{2\nu(r)}, e^{-2\lambda(r)}, -r^2, -r^2\sin^2\theta)$, $\nu(r)$ e $\lambda(r)$ being functions to be obtained from the solution of Einstein's equations.

Under the standard assumption that matter in the star interior behaves as an ideal fluid, the energy-momentum tensor can be written in the form

$$T_{\mu\nu} = [\epsilon(r) + P(r)]u_\mu u_\nu - P(r) g_{\mu\nu}\,, \tag{6.33}$$

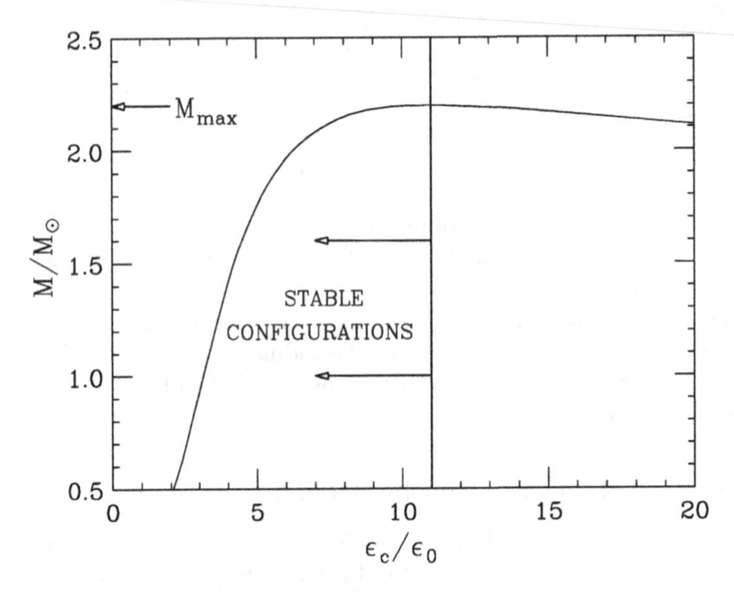

Figure 6.4 Neutron star mass obtained from the solution of the TOV equations, displayed as a function of the central energy density in units of $\epsilon_0 = 140$ MeV/fm^3. The calculation has been performed using the EOS of β-stable nuclear matter referred to as APR2 model [121].

where $\epsilon(r)$ and $P(r)$ denote the space distribution of mass-energy density and pressure, respectively, while $u_\mu = \partial x^\mu/\partial\tau$, τ being the proper time, is the local four-velocity. In the case of non rotating stars $u^\mu \equiv (e^{-\nu(r)}, 0, 0, 0)$, and the energy-momentum tensor turns out to be diagonal.

The Einstein tensor $G_{\mu\nu}$ is defined as

$$G_{\mu\nu} = R_{\mu\nu} - \frac{1}{2} g_{\mu\nu} R \ , \tag{6.34}$$

where $R_{\mu\nu}$ and R can be written in terms of Christoffel symbols [136].

From the tt component of Einstein's equation we obtain the relation

$$e^{-\lambda(r)} = 1 - \frac{2}{r} M(r) \ , \tag{6.35}$$

with

$$M(r) = 4\pi \int_0^r \epsilon(r') r'^2 dr' \ , \tag{6.36}$$

while the rr component yields

$$\frac{d\nu}{dr} = -2 \frac{4\pi P(r) + \dfrac{M(r)}{r^2}}{1 - \dfrac{2}{r} M(r)} \ . \tag{6.37}$$

Combining Eqs.(6.35), (6.36) and (6.37) with the requirement of energy-momentum conservation, implying

$$\frac{d\nu}{dr} = -2 \frac{2}{\epsilon(r) + P(r)} \frac{dP}{dr} , \tag{6.38}$$

we finally obtain the system of differential equations

$$\frac{dP(r)}{dr} = -\epsilon(r) \frac{GM(r)}{r^2} \left[1 + \frac{P(r)}{\epsilon(r)} \right] \left[1 + \frac{4\pi r^3 P(r)}{M(r)} \right] \left[1 - \frac{2GM(r)}{r} \right]^{-1} , \tag{6.39}$$

$$\frac{dM(r)}{dr} = 4\pi r^2 \epsilon(r) , \tag{6.40}$$

which was first derived by R. Tolman, J. Oppenheimer and G. Volkov [6, 137] in 1939.

The first term in the right-hand side of Eq.(6.39) is the Newtonian gravitational force, similar to the one appearing in Eq.(6.27) but with matter density replaced by mass-energy density. The first two additional factors take into account relativistic corrections, that become vanishingly small in the limit $k_F/m \to 0$, m and k_F being the mass and Fermi momentum of the star constituents, respectively. Finally, the third factor describes the effect of space-time curvature. Obviously, in the non relativistic limit Eqs.(6.39) and (6.40), referred to as Tolman-Oppenheimer-Volkof (TOV) equations, reduce to the classical equilibrium equation (6.27) and (6.28).

For any values of the mass-energy density at the centre of the star, $\epsilon_c = \epsilon(0)$, the TOV equations determine the mass and radius of the corresponding equilibrium configuration. The value of the maximum mass is associated with the mass-energy density $\bar{\epsilon}_c$ such that

$$\left(\frac{dM}{d\epsilon_c} \right)_{\epsilon_c = \bar{\epsilon}_c} = 0 , \tag{6.41}$$

and stable configurations have $dM/d\epsilon_c > 0$. In the region in which $dM/d\epsilon_c < 0$ equilibrium configurations are unstable with respect to radial oscillations.

The ϵ_c dependence of the neutron star masses obtained using the APR2 model of the nuclear matter EOS is illustrated in Fig. 6.4.

CHAPTER 7

CONSTRAINTS FROM ASTROPHYSICAL DATA

Neutron stars have long been recognised as a potentially unparalleled source of information on strong interactions dynamics in the regime corresponding to high density and low temperature, which is not accessible in terrestrial laboratories. In principle, the data collected from neutron star observations, the extent of which has been dramatically extended by the first detection of gravitational waves in the 2010s, may in fact set stringent constraints on theoretical models of neutron star matter.

While many ongoing studies are primarily aimed at exploiting the available and forthcoming empirical information to determine the occurrence of phase transitions—leading to the appearance of exotic states of matter featuring strange baryons or deconfined quarks, possibly in a colour superconducting phase, in the inner core of the star—our discussion will be limited to the regime in which nucleons are believed to be the relevant degrees of freedom, and nuclear matter theory is expected to be applicable.

The basic tenet underlying the study of neutron star properties within nuclear matter theory is that the microscopic dynamics is completely decoupled from gravitational effects, which are known to play a critical role in determining neutron star properties. This assumption is justified by the observation that, in the density region typical of neutron stars, the range of nuclear forces is negligible compared to the radius of curvature of the space. As a consequence, NN interactions can be treated as if they were taking place in flat space. In their classic book, Harrison *et al.* estimate the density at which the validity of this approximation breaks down to be $\varrho \sim 10^{49}$ g cm^{-3} [138].

7.1 MEASUREMENTS OF MASS AND RADIUS

The connection between the EOS of neutron star matter and the star mass clearly emerges from the results of the pioneering study carried out by J. Oppenheimer and G. Volkoff [6] in the 1930s, showing that the mass of a star consisting of non interacting neutrons is nearly constant for central densities $\varrho_c \gtrsim 2\varrho_0$, and cannot exceed $\sim 0.8\ M_\odot$. This result unambiguously shows that the Fermi gas EOS is extremely soft, and fails to support neutron stars with masses compatible with observations.

The recognition that the pressure responsible for neutron star equilibrium cannot be explained by Fermi-Dirac statistics alone, but is primarily of dynamical origin, has important phenomenological implications. It strongly suggests that experimental information on neutron star masses may be used to test the predictions of theoretical models of nuclear dynamics.

The masses of neutron stars belonging to a binary system have been determined to remarkable accuracy from the measurement of the keplerian orbital parameters, entering the definition of the so-called mass functions [139]. More recently, these analyses have been refined, by including the experimental information on post-keplerian parameters, describing effects of general relativity such as the advance of periastron and the gravitational red-shift [140]. This approach allowed the precise determination of the masses of the neutron stars in the binary system PSR1913+16, discovered by R. Hulse and J. Taylor in 1975 [141]. This analysis yielded the results $M_1 = 1.39 \pm 0.15 M_\odot$ and $M_2 = 1.44 \pm 0.15 M_\odot$ [142].

Because to each EOS is associated a maximum mass M_{max}, see Section 6.4, the most straightforward test of the EOS is a comparison between M_{max} and the values of the measured neutron star masses, summarised in Fig. 7.1.

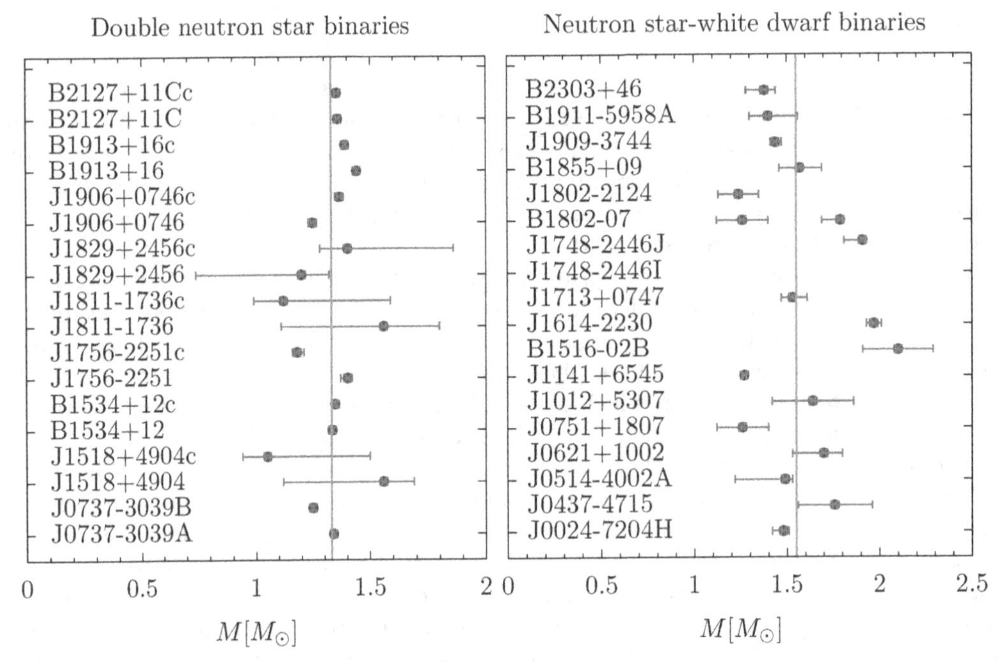

Figure 7.1 Distribution of measured neutron star masses, obtained from analyses of double neutron star binaries (left panel) and neutron star-wihte dwarf binaries (right panel). The central values of the mass distributions, corresponding to $M = 1.33$ and $1.55\ M_\odot$, respectively, are represented by the vertical lines. Data taken from [143]. .

It is apparent that, although the distribution is peaked around the canonical mass $1.4\ M_\odot$, the data point corresponding to the pulsar J1614-2230, with a mass $M = 1.97 \pm 0.04\ M_\odot$ precisely determined in 2010 from the observation of the Shapiro delay [144], strikingly sticks out. Another measurement of a neutron star mass around two solar masses, not included in the figure, has been reported in 2013 [145].

Such large masses appear to rule out models predicting a soft EOS, most notably those featuring the onset of a high-density phase including hyperons. Unfortunately, however, little is known about hyperon dynamics, and these models often imply crude assumptions.

On the other hand, it is a fact that many models based only on nucleon degrees of freedom turn out to be compatible with the recent measurements.

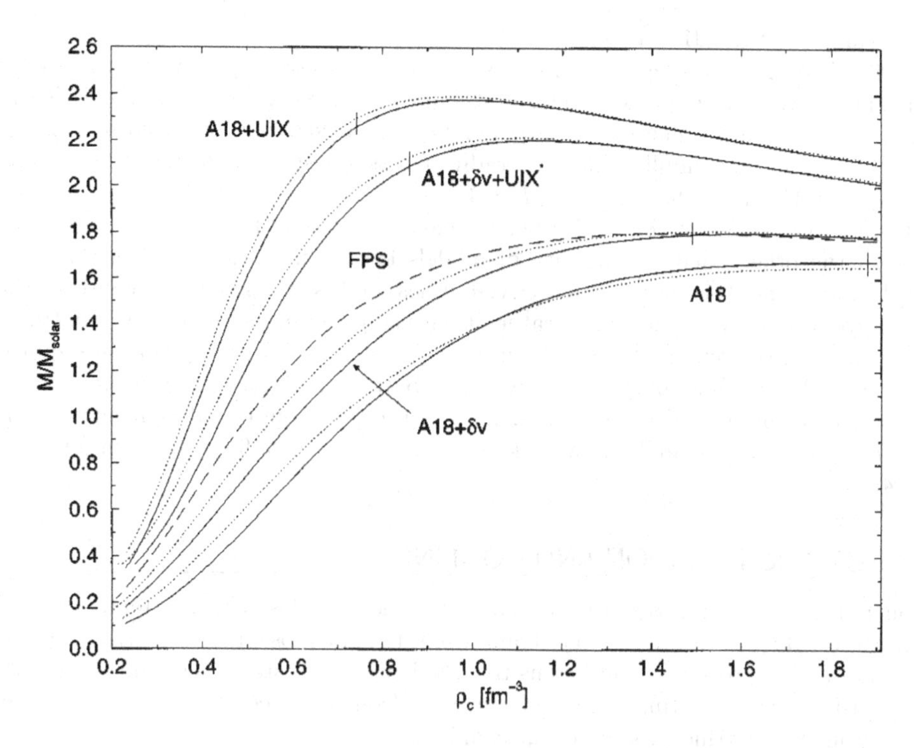

Figure 7.2 Neutron star masses, expressed in units of the solar mass, as a function of central density. The curves have been obtained using different nuclear Hamiltonians. The meaning of the labels is the same as in Figs. 5.13 and 5.14. Taken from [121].

As an example of the results obtained from nuclear matter theory, Fig. 7.2 shows the neutron star masses obtained from different models of EOS, as a function of central density [121]. The meaning of the labels is same as in Fig. 5.14.

It clearly appears that the inclusion of NNN interactions, described using the UIX and UIX′ models, is required to generate the pressure needed to support a mass $\gtrsim 2\ M_\odot$. However, as pointed out in Section 3.2.1, the stiffness of the EOS leads to a violation of causality at some large density $\bar{\varrho}$. The values of $\bar{\varrho}$ corresponding to the AV18+UIX and APR2 models are indicated by the vertical bars of Fig. 5.14.

The inclusion of boost corrections appears to alleviate the violation of causality, pushing $\bar{\varrho}$ towards higher values. Note that the values of $\bar{\varrho}$ corresponding to both models including NNN interactions turn out to be larger than the central density of a neutron star having $M \sim 2\ M_\odot$.

A more severe constraint may be obtained from measurements of the neutron star radii. However, the precision of the available data does not allow to impose stringent conditions on theoretical models. It has been suggested that valuable complementary information may come from measurements of the gravitational redshift, z, related to the M/R ratio through

$$R(1+z) = R\left(1 - 2G\frac{M}{R}\right)^{-1/2} , \tag{7.1}$$

with G being the gravitational constant.

In 2002, Cottam *et al.* reported the observation of spectroscopic lines associated with oxygen and iron transitions in the spectra of 28 bursts of the x-ray binary system EXO07948-676 [146]. The redshift value extracted from their analysis, $z = 0.35$, corresponding to $M/R = 0.153\ M_{\odot}/\text{km}$, implies that the radii of neutron stars with masses in the range $1.4 \lesssim M \lesssim 1.8\ M_{\odot}$ lie in the range $9 \lesssim R \lesssim 12$ km.

Although the results of this and other similar studies tend to rule out soft EOS, supporting on the other hand the validity of models based on Nuclear Matter Theory, the interpretation of the data involves non trivial issues, and is still somewhat controversial.

It has to be also mentioned that, rather than using the measurements of neutron star radii to constrain the nuclear EOS at high density, one may follow the opposite approach, and try to use the reliable information on the EOS at subnuclear density to determine the radius of a neutron star of known mass. This is the philosophy underlying the work by Hebeler *et al* [147], whose analysis predicts a radius in the range 9.7 – 13.9 km for a star of mass 1.4 $M_{\odot}$.

7.2 NEUTRINO EMISSION AND COOLING

Neutron stars are formed in the aftermath of a supernova explosion with a temperature $T \sim 10^{11}$ K. After ~ 30 sec the star becomes transparent to neutrinos and antineutrinos produced in weak interaction processes, and begins to cool. During this stage, the temperature drops to $T \sim 10^8$ K within a timescale that can vary from few weeks to thousands of years, depending on the neutrino production mechanism.

The dominant processes leading to neutrino and antineutrino emission in β-stable nuclear matter are driven by the charged weak current. The simplest mechanisms of this kind are neutron β-decay, depicted in the left panel of Fig. 7.3, and electron capture by a proton

$$n \to p + e + \bar{\nu}_e , \tag{7.2}$$
$$p + e \to n + \nu_e . \tag{7.3}$$

The above reactions, first discussed by Gamow and Schoenberg in the 1940s [148, 149], are referred to as direct Urca processes.

Because matter is assumed to be transparent to neutrinos, chemical equilibrium with respect to the rates of processes (7.2) and (7.3) requires that

$$\mu_n = \mu_p + \mu_e , \tag{7.4}$$

with μ_α being the chemical potential of particles of type α, see Section 6.2.2.

The calculation of the emissivity associated with the direct Urca process—defined as the amount of energy carried away by neutrinos and antineutrinos per unit time and unit volume—is computed exploiting the near degeneracy of neutrons, protons and electrons.

Neutrinos and antineutrinos give identical contributions to the emissivity, Q_D, which turns out to depend on the dynamical model underlying the EOS of neutron star matter through the neutron and proton effective masses, defined in Eq. (4.9), and exhibits a T-dependence determined by the phase space available to the final state particles, see, e.g.,

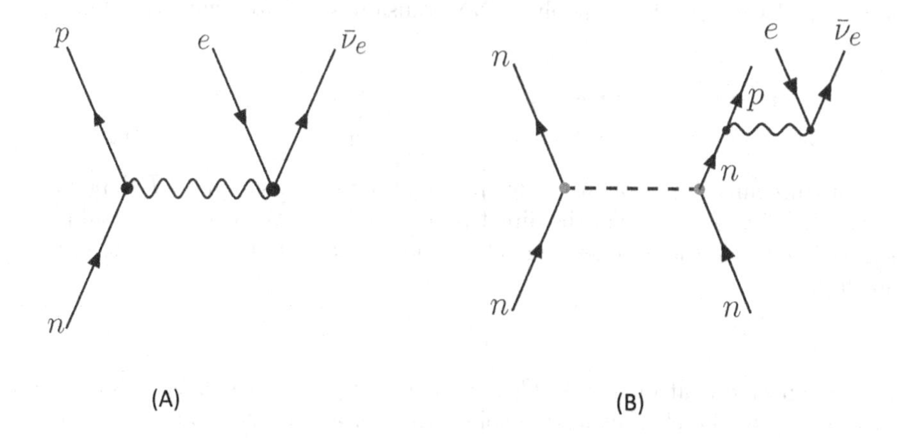

Figure 7.3 Panel (A): diagrammatic representation of antineutrino emission in neutron β-decay, see Eq. (7.2). Panel (B): diagrammatic representation of antineutrino emission in the aftermath of a neutron-neutron scattering process, see Eq. (7.8). The wiggly line represents the exchange of a W boson, while the dashed line depicts the neutron-neutron interaction.

Ref. [139]. We find the result

$$Q_D \propto \varrho_e^{1/3} \frac{m_p^\star m_n^\star}{m^2} T_9^6 \Theta_{npe} \ , \tag{7.5}$$

where T_9 denotes the temperature expressed in units of 10^9 K, ϱ_e is the electron density, $m_p^\star$ and $m_n^\star$ are the proton and neutron effective masses, respectively, and

$$\Theta_{npe} = \begin{cases} 1 & \text{if } p_{Fn} \leq p_{Fp} + p_{Fe} \\ 0 & \text{otherwise} \end{cases} \ , \tag{7.6}$$

with $p_{F\alpha}$ being the Fermi momentum of particles of type α.

The above definition implies the existence of a threshold for the direct Urca process, whose occurrence is determined by the proton fraction x. To see this, consider that, because charge neutrality requires $\varrho_e = \varrho_p$, that is, $p_{F_e} = p_{F_p}$, the proton fraction must fulfil the relation

$$x = \frac{\varrho_p}{\varrho_p + \varrho_n} \geq \frac{1}{9} \ . \tag{7.7}$$

A similar threshold condition is obtained in the case of weak interaction processes involving both electrons and muons.

Note that, in the absence of phase transitions, the proton fraction of nuclear matter, x, turns out to be a monotonically increasing function of the density, ϱ. Therefore, the threshold condition on x translates into a corresponding condition on ϱ, which in turns implies the existence of a threshold neutron star mass, $M_{\rm thr}$. The direct Urca process can only occur in stars with mass $M \geq M_{\rm thr}$.

If the direct Urca process is not allowed, the most efficient mechanism of neutrino emission is the modified Urca process, in which momentum conservation is made possible by the presence of an additional spectator nucleon.

The modified Urca reactions involve a NN collision associated with the β-decay or capture reaction

$$n+n \to n+p+e+\bar{\nu}_e \quad , \qquad n+p+e \to n+n+\nu_e \ , \tag{7.8}$$

$$p+n \to p+p+e+\bar{\nu}_e \quad , \qquad p+p+e \to p+n+\nu_e \ . \tag{7.9}$$

The neutrino emissivities of the above processes exhibit a common T-dependence, that differs from the T_9^6 dependence of the direct processes due to the two additional phase space factors associated with the degenerate nucleons not involved in the weak interaction. In this case, one finds

$$Q_M \propto T_9^8 \ . \tag{7.10}$$

Knowing the specific heat of matter C_V, the emissivity, which will be denoted as Q, can be used to estimate the characteristic cooling time of a star, τ. From the thermal energy balance

$$Q = \frac{dE_{\rm th}}{dt} = -C_V \frac{dT}{dt} \ , \tag{7.11}$$

it follows that

$$\frac{1}{\tau} = \frac{Q}{C_V T} \ . \tag{7.12}$$

The value of τ turns out to be much shorter—weeks compared to thousands of years—if the direct Urca processes are active [150]. For this reason, the scenarios in which direct and modified Urca processes are the dominant neutrino emission mechanisms are referred to as fast and slow cooling, respectively[1].

The possible occurrence of an intermediate scenario, referred to as medium cooling [152], is associated with the onset of neutron superfluidity at temperatures below a critical value T_c, leading to increased neutrino emission due to the formation and dissociation of Cooper pairs.

The medium cooling scenario with critical temperature $T_c = 5.5 \times 10^8$ K provides a remarkably good description of the observed time evolution of the temperature of the neutron star in Cassiopea A [153, 154], see Fig. 7.4. The vertical axis of the figure gives $T_e^\infty = (1+z)T_e$, where z is the redsfhift and the effective temperature T_e is defined so that the total photon luminosity of the star can be written using the standard blackbody emission formula. Note that data included in the figure are the only available measurement of neutron star cooling. The theoretical results correspond to a star of mass $M = 1.4 M_\odot$, in which the direct Urca process is forbidden.

7.3 GRAVITATIONAL-WAVE OBSERVATIONS

On August 17, 2017, the Advanced LIGO-Virgo detector network performed the first observation of a gravitational wave (GW) signal—labelled GW170817—that turned out to be consistent with emission from the inspiral of a binary neutron-star system [156] . The detection of this signal, supplemented by the later observation of electromagnetic radiation by space- and and ground-based telescopes [157], can be seen as the dawning of the long-anticipated age of gravitational-wave astronomy.

[1] In his autobiography, Gamow reports that the direct Urca process is named after the Urca Casino—in Rio de Janeiro, Brazil—where money disappeared from the pockets of the gamblers as rapidly as energy is carried away from the stars by neutrino emission [151].

The information provided by GW data will allow a significant reduction of the uncertainties on the measured equilibrium properties of neutron stars, primarily mass and radius, thus putting further constraints on the dynamical models underlying the EOS of matter in the star interior. In addition, the detection of GW originating from excitation of quasi-normal modes (QNM), is expected to shed light on non-equilibrium properties, such as the transport coefficients associated with thermal and electric conductivity and viscosity, whose consistent description within nuclear mater theory involves non trivial issues.

7.3.1 Neutron star merger

Gravitational waves emitted during the inspiral leading to a binary neutron star merger are driven by the tidal deformation of the participating stars, which is largely determined by the EOS, see [158] and references therein.

Many theoretical studies are carried out using models of the EOS which are only partially derived from a microscopic description of the dynamics of dense nuclear matter [159], or obtained from simple phenomenological parametrisations, based on the information provided by measured nuclear properties [160]. Even though the results of these analyses provide valuable information and are useful for the interpretation of the data, the use of fully microscopic models of the EOS, derived within the framework of nuclear matter theory and applicable over the whole range of densities relevant to neutron stars, is needed to fully exploit the potential of gravitational-wave observations.

The orbital motion of a binary neutron star system is associated with the emission of

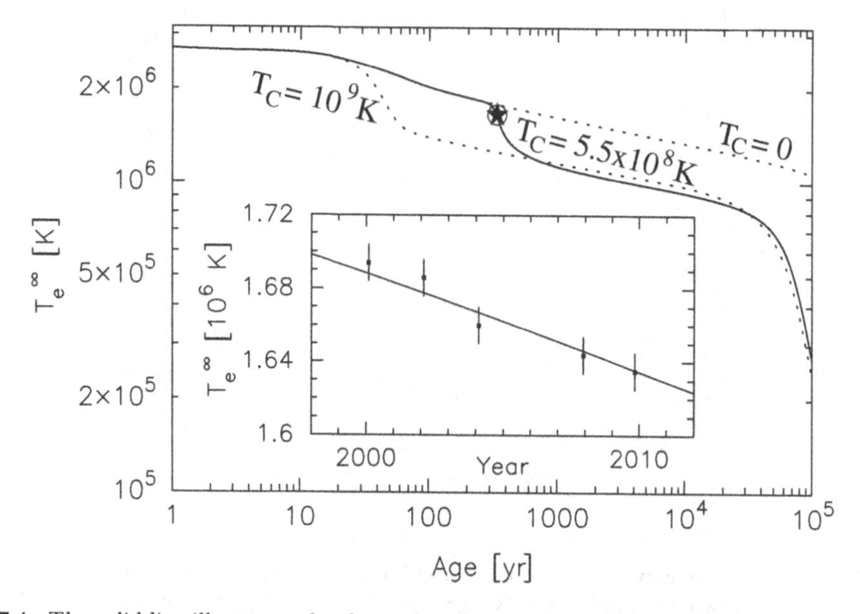

Figure 7.4 The solid line illustrates the thermal evolution obtained within the intermediate cooling scenario setting the star mass to 1.4 $M_\odot$ and the critical temperature to $T_c = 5.5 \times 10^8$ K. For comparison, the dotted lines display the behaviour corresponding to $T_c = 0$, that is, in the absence of neutron superfluidity, and 10^9 K, respectively. The inset shows a comparison with the observed evolution of the neutron star in Cassiopea A [153, 154]. Taken from [155]

GW carrying away energy and angular momentum. This process leads to a decrease of the orbital radius and, conversely, to an increase of the orbital frequency.

In the early stage of the inspiral, characterised by large orbital separation and low frequency, the two stars—of mass m_1 and m_2, with $m_1 \geq m_2$—behave as point-like bodies and the evolution of the frequency is primarily determined by the chirp mass $\mathcal{M}$, defined as

$$\mathcal{M} = \frac{(m_1 m_2)^{3/5}}{(m_1 + m_2)^{1/5}} . \tag{7.13}$$

The chirp mass of event GW170817, measured to remarkable accuracy, turns out to be $\mathcal{M} = 1.188^{+0.004}_{-0.002} M_\odot$. On the other hand, the determination of the masses of the components depends on the assumptions made on their spins. The data have been analysed for two different scenarios, corresponding to high and low spin. The results, yielding the range of both masses at 90% confidence level, are $m_1 \in \{1.36, 2.26\} M_\odot$ and $m_2 \in \{0.86, 1.36\} M_\odot$ for the high-spin scenario, and $m_1 \in (1.36, 1.60) M_\odot$ and $m_2 \in (1.17, 1.36) M_\odot$ for the low-spin scenario. Note that the high-spin scenario is compatible with the recently observed neutron stars masses around two solar masses.

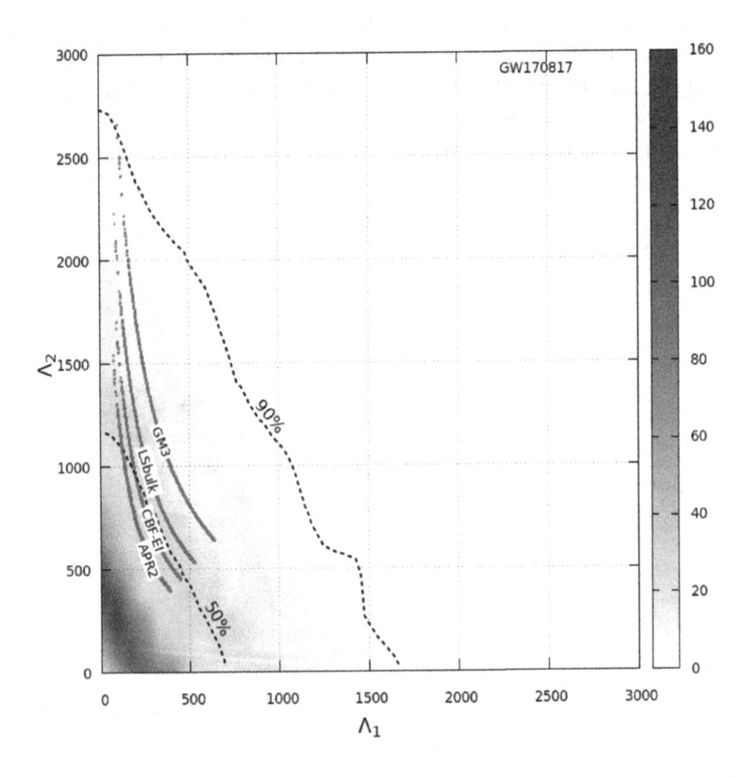

Figure 7.5 Probability density for the tidal deformability parameters inferred by the detected signals of the inspiral of two neutron stars, GW17081027. Theoretical results are labelled as described in the text. The dashed lines represent the contours enclosing 50% and 90% of the posterior probability density. Taken from [161].

The details of the internal structure of the stars begin to be important as soon as the orbital separation becomes comparable to the stellar radius. In this regime, the tidal field of one of the neutron stars induces a mass-quadrupole moment on the companion, which in turn generates the same effect on the first star, thus accelerating coalescence. This effect is quantified by a parameter describing tidal deformability, defined as

$$\Lambda_i = \frac{2}{3} k_2 \left(\frac{R_i}{G m_i} \right)^5 , \tag{7.14}$$

where G is the gravitational constant, m_i and R_i, with $i = 1, 2$, are the mass and radius of the star, and k_2 is referred to as second tidal Love number [162]. For any given mass, the radius and the tidal Love number are uniquely determined by the EOS of neutron star matter.

Figure 7.5 illustrates the dependence of tidal deformability on the the EOS, assumed to be the same for the two stars. The values of Λ_1 and Λ_2 have been computed from the Love numbers predicted by different EOS, using the observed mass distributions.

The theoretical results are compared to the probability density of Λ_1 and Λ_2 reported by the LIGO-Virgo collaboration for the high-spin scenario. The labels APR2, CBF-EI and GM3 correspond to the APR2 and CBF effective interaction models described in Section 5.4, and to an extension of the Walecka model of Section 4.7.3, which includes the contribution of the vector-isovector ρ-meson [163]. The LS-bulk EOS, specifically designed for easy implementation in stellar collapse simulations, has been obtained from a non relativistic momentum-dependent effective interaction constrained by nuclear phenomenology [164]. The comparison shows a non negligible sensitivity to the dynamical model determining the EOS, and suggest that models more constrained by phenomenology provide a better account of the observations.

From the above discussion, it follows that the information on tidal deformability and mass may be used to obtain an estimate of the star radius. A comparison between the mass-radius relations obtained from different models of EOS and the range of mass and radius compatible with the observations is shown in Fig. 7.6. The emerging pattern is consistent with that of Fig. 7.5.

7.3.2 Quasi-normal Modes

When a neutron star is perturbed by some external or internal event, it can be set into non radial oscillations, associated with emission of gravitational waves at the characteristic frequencies of its quasi-normal modes. This may happen, for example, as a consequence of a glitch, a close interaction with an orbital companion, a phase transition occurring in the inner core, or in the aftermath of a gravitational collapse. The frequencies and damping times of the quasi-normal modes carry information on the structure and dynamics of the star, and their observation will shed light on a variety of properties of matter in its interior.

Relation to the EOS

Quasi-normal modes are classified according to the source of the restoring force which prevails in bringing the perturbed element of fluid back to the equilibrium position. For example, in the case of g-modes , the restoring force is mainly provided by buoyancy, while for a p-modes it is due to a gradient of pressure. The frequencies of the g-modes are lower than those of the p-modes, and the two sets are separated by the frequency of the fundamental f-mode , associated with global oscillations of the fluid. General relativity predicts the existence of additional modes, called w-modes, which are purely gravitational, and do not induce any fluid motions.The w-modes can be both polar or axial, depending on parity. They are strongly damped and, in general, their frequencies are higher than those of the p-modes.

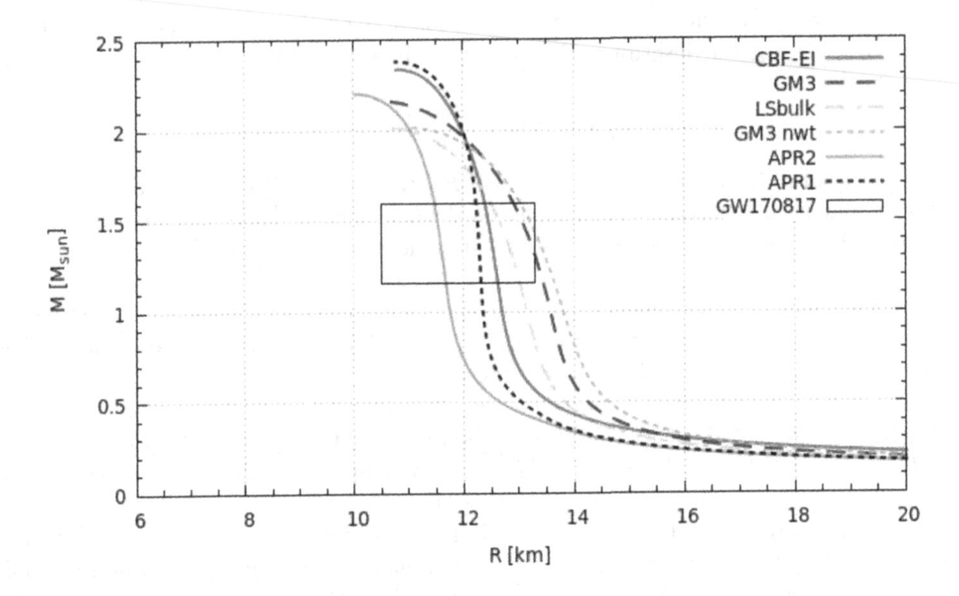

Figure 7.6 Mass-radius diagram for different models of EOS. The region within the black square represents the 90% confidence level result reported by the LIGO-Virgo Collaboration [157]. Taken from [161].

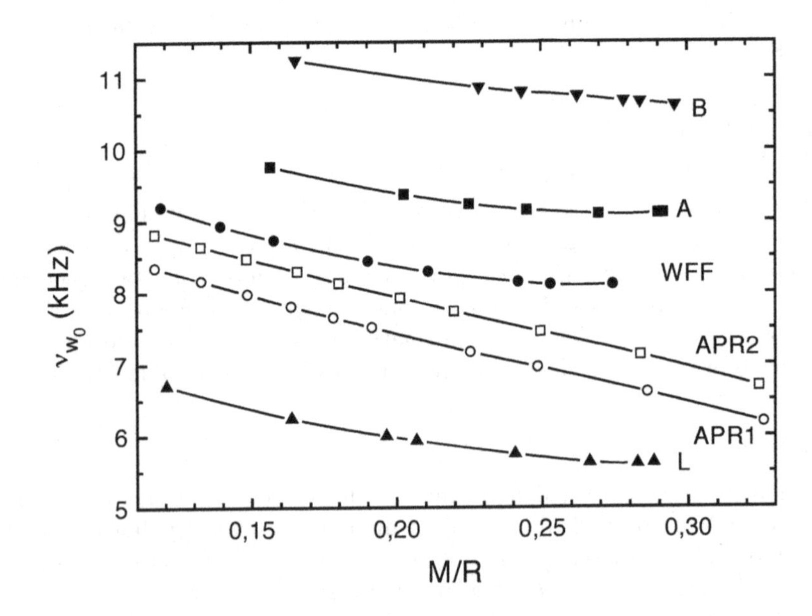

Figure 7.7 Frequency of the first axial w-mode obtained from different models of EOS, plotted as a function of the source compactness. Taken from [165].

To see an example of how the oscillation frequencies of QNMs may depend on the EOS of neutron star matter, let us consider the axial w-modes of a non rotating star. The corresponding complex frequencies are eigenvalues of a Schrödinger-like equation featuring a potential $V_\ell(r)$ that explicitly depends upon the EOS according to

$$V_\ell(r) = \frac{e^{2\nu(r)}}{r^3} \left\{ \ell(\ell+1)r + r^3 \left[\epsilon(r) - P(r)\right] - 6M(r) \right\} , \tag{7.15}$$

with $\nu(r)$ and $M(r)$ defined as in Section 6.4.1, and

$$\frac{d\nu}{dr} = -\frac{1}{[\epsilon(r) + P(r)]} \frac{dP}{dr} . \tag{7.16}$$

In the above equations, P and ϵ are the energy density and pressure of neutron star matter, assumed to behave as a perfect fluid. Given a model of the EOS, the complex eigenvalues, ω, are uniquely determined.

As an example, Fig. 7.7 shows the frequencies—that is, the real parts of the eigenvalues—corresponding to a somewhat outdated, but still representative, set of EOS, plotted as function of the star compactness M/R [165].

The pattern emerging from Fig. 7.7 strictly reflects the stiffness of the different EOS in the relevant density region (typically $\varrho_0 < \varrho < 5\varrho_0$), with softer EOS corresponding to higher frequencies. For example, the curve labelled B has been obtained from a model including hyperons, whose appearance leads to a very soft EOS and a maximum mass $\sim 1.4\ M_\odot$. On the other hand, the curve labelled L corresponds to a relativistic mean-field model involving nucleon degrees of freedom only, yielding a maximum mass $\sim 2.7\ M_\odot$.

The frequencies turn out to depend rather weakly on compactness, and the corresponding curves do not cross one another over the range displayed in the figure. Therefore, in principle a measurement of ν_{W_0} may allow to discriminate between different models of neutron star matter and pin down, for example, the occurrence of strange baryonic matter in a star of known mass, independent of its radius.

Towards GW asteroseismology

In the late 1990s, N. Andersson and K. Kokkotas [166] computed the frequencies of the f-mode, the first p-mode and the first polar w-mode of a non rotating neutron star for a number of EOS available at that time. They fitted the results of their calculations with appropriate universal functions of the macroscopic properties of the star, that is, mass and radius, and showed how the empirical relations could be used to put constraints on these quantities if the frequency of one or more modes could be identified in a detected gravitational signal.

A similar analysis has been carried in the 2000s using an updated set of EOS, including state-of-the-art results of nuclear matter theory [167]. As an example, Fig. 7.8 shows the frequencies of the fundamental f-mode corresponding to different EOS, plotted as as a function of the square root of the average density. It appears that the frequencies corresponding to different EOS tend to scale to a straight line. The sizeable displacement, of ~ 100 Hz, of the results obtained from the revisited analysis reflects more than a decade of progrerss in theoretical modelling of the EOS.

The damping time of the f-mode and the frequencies and damping times of the p- and w-modes also exhibit scaling, when plotted as a function of the compactness M/R. As a consequence, the identification of the frequency and damping time of some of these modes in a detected gravitational signal would allow the determination of the radius of the source, knowing its mass.

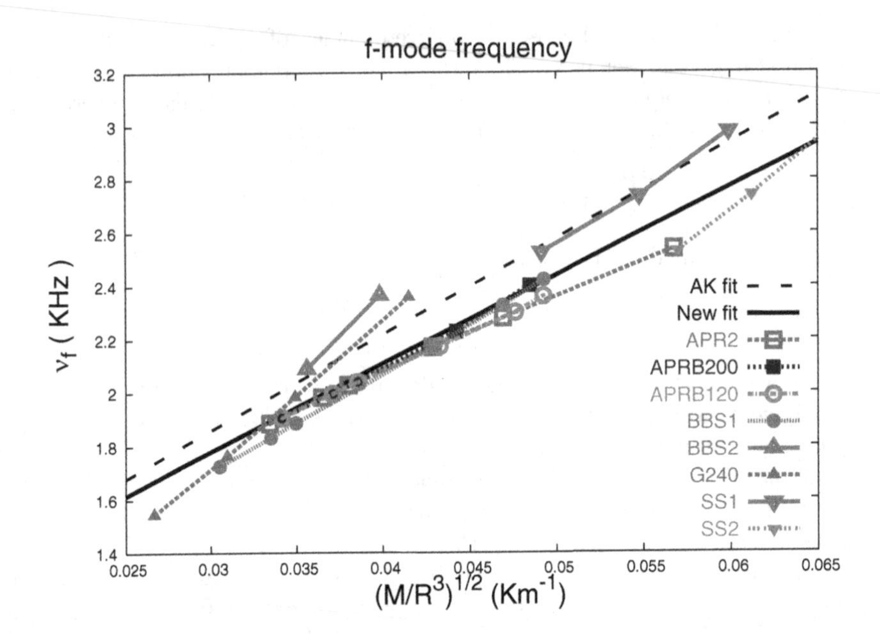

Figure 7.8 Frequency of the f-mode plotted as a function of the square root of the average density for different models of the EOS. The fits reported by the authors of Refs. [166] and [167] are labelled AK fit and New fit, respectively. Taken from [167].

r-mode instability of rotating stars

As S. Chandrasekhar first pointed out at the beginning of the 1970s [168, 169], GW emission following the excitation of QNMs—which in non rotating stars is a dissipative process causing the damping of oscillations—may instead lead to the instability of rotating stars. At the end of the same decade, J. Friedman and B. Schutz [170] proved that, owing to the mechanism discussed by Chandrasekhar, all rotating stars are in fact unstable if matter in their interior behaves as a perfect fluid. Finally, in the 1990s it has been demonstrated that the so called r-modes, that is, oscillations of rotating stars whose restoring force is the Coriolis force , are driven unstable by GW emission in all perfect fluid stars [171, 172]. If, on the other hand, neutron star matter is not a perfect fluid, dissipative mechanisms, such as viscosity, can damp the modes leading to the Chandrasekhar-Friedman-Schutz (CFS) instability, or even suppress them altogether.

The onset of the r-mode instability is regarded as a prominent mechanism associated with the emission of potentially detectable GW [173], although the determination of the stability limits poses a severe challenge to microscopic models of neutron star matter.

In addition to the EOS, theoretical analyses require the description of the viscosity of matter in the star interior, which in turn depends on the possible occurrence of superfluid and/or superconducting phases. The EOS of both the normal and superfluid/superconducting phases is generally obtained from accurate calculations, carried out within the framework of nuclear matter theory using phenomenological Hamiltonians, strongly constrained by nuclear systematics and nucleon-nucleon scattering data. On the other hand, theoretical studies of transport properties, involving the calculation of the collision integral appearing in the linearised Boltzmann equation, have often resorted to oversimplified models of nuclear dynamics.

A step towards a fully consistent determination of the EOS and the shear viscosity coefficient, η, has been made in the 2000s by the authors of Refs. [174, 175]. In these studies, the calculation of η has been carried out using the in-medium neutron-neutron scattering cross section obtained from the density-dependent CBF effective interaction described in Section 5.4, which was also consistently employed to compute the EOS. The density dependence of the differential cross section at $E_{\rm cm} = 100$ MeV, displayed as a function of the scattering angle in the center-of-mass frame, is illustrated in Fig. 7.9.

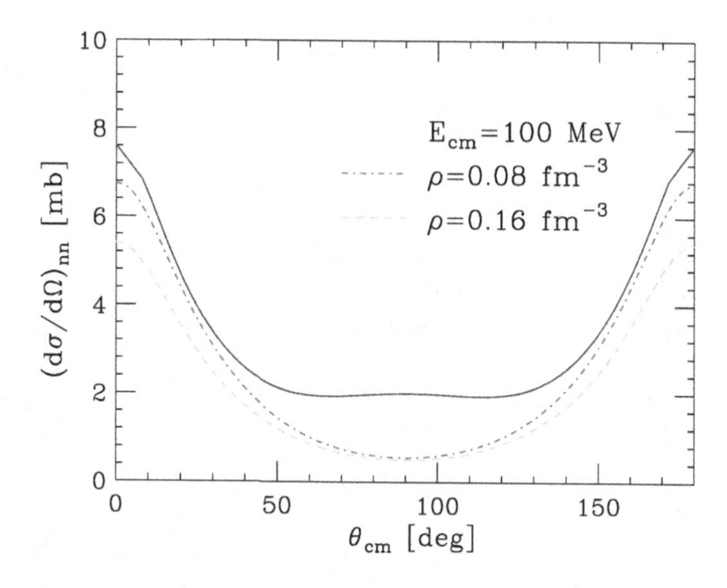

Figure 7.9 Differential neutron-neutron scattering cross section in PNM at $E_{\rm cm} = 100$ MeV, as a function of the scattering angle in the center-of-mass frame. The dashed and dot-dashed lines have been computed using the CBF effective interaction described in Section 5.4 at density $\varrho = 0.16$ and 0.08 fm^{-3}, respectively. For comparison, the solid line shows the results obtained neglecting medium effects. Taken from [174].

Figure 7.10 shows the T-independent quantity ηT^2, with T being the temperature, evaluated for PNM using an Hamiltonian including only the AV6P NN potential. For comparison, the figure also includes results obtained from G-matrix perturbation theory with the same NN potential. The effects of medium modifications of the NN cross sections can be gauged comparing the solid and dot-dash lines. The latter corresponds to a calculation in which these effects have been neglected altogether, following an approximation scheme widely employed in early studies of the viscosity of nuclear matter.

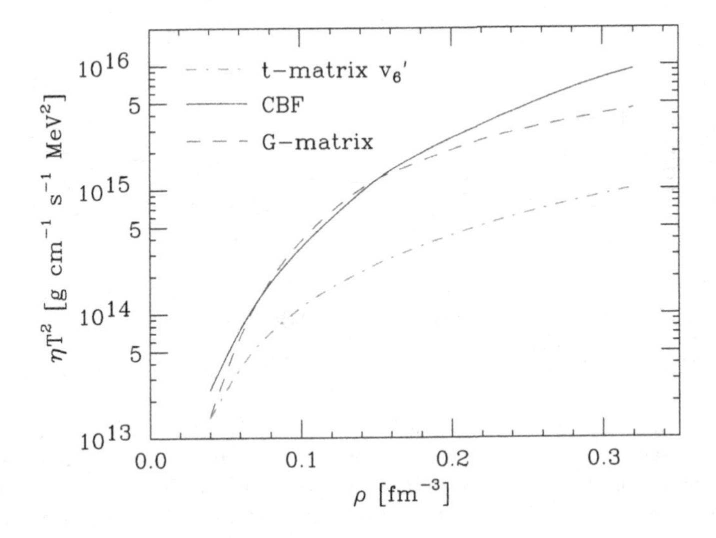

Figure 7.10 Density dependence of the quantity ηT^2, η and T being the shear viscosity coefficient of PNM and the temperature. The solid and dashed lines correspond to the results of the CBF effective interaction and G-matrix approaches, respectively, while the dot-dash line shows the results obtained from the free-space t-matrix. All calculations have been performed using the AV6P potential. Taken from [175].

Outlook

Almost half a century after publication of H. Bethe's seminal paper of 1971, nuclear matter theory continues to be a most active field of research, encompassing the derivation of realistic models of nuclear dynamics, as well as the development of accurate computational schemes for the solution of the quantum mechanical many-body problem.

State-of-the-art phenomenological potentials provide an accurate description of the properties of the two-nucleon system, in both bound and scattering states, over the broad energy range relevant to neutron star applications. The nuclear Hamiltonians constructed supplementing these potentials with additional contributions describing irreducible three-nucleon interactions—needed to explain the binding energies of the three-nucleon systems as well as saturation of isospin-symmetric nuclear matter—have been shown to possess a remarkable predictive power. An important and still open issue of the phenomenological approach is the consistent and unambiguous determination of the three-body potential, which will also require a careful analysis of the corrections to the non relativistic approximation.

The development of a new generation of nuclear Hamiltonians within the framework of chiral effective field theory has allowed to establish a link between nuclear dynamics and the fundamental theory of strong interactions. However, the low momentum expansion laying at the basis of this formalism, while being adequate to obtain an accurate description of nuclear properties, presently limits its nuclear matter applications to densities lower than twice the nuclear saturation density, thus excluding the region of the neutron star core.

As far as computational techniques are concerned, the development of Quantum Monte Carlo methods has achieved the level required to carry out calculations of the zero-temperature equation of state of pure neutron matter using realistic Hamiltonian, and the extension to isospin-symmetric matter is certainly within reach of the available algorithms and computing capability.

More flexible schemes—such as the variational method underlying correlated basis function perturbation theory and the self consistent Green's function method—have allowed to extend nuclear matter calculations to fundamental properties other that the ground-state energy, such as the nucleon spectral functions and momentum distributions. The results of these studies have been extensively employed to analyse the measured electron-nucleus scattering cross sections in a broad kinematical range. Moreover, currently being exploited to develop a framework for the interpretation of the signals detected by long-baseline neutrino experiments.

Besides being highly valuable in their own right, the results of accurate Monte Carlo studies have also provided the benchmarks needed to assess the accuracy of variational estimates of the energy of nuclear matter, which are in turn used as an input for the determination of microscopic effective interactions within the framework based on correlated wave functions and the cluster expansion technique. This procedure, which allows to take into account the density-dependent screening of nuclear interactions due to the presence of the nuclear medium, may be seen as a coordinate-space implementation of renormalisation-group evolution. The resulting effective interaction is well behaved, and can be used to carry out perturbative calculations in the Fermi gas basis, thus allowing for a consistent treatment of equilibrium and non equilibrium properties of nuclear matter at arbitrary proton fraction and spin polarisation, for both zero and non-zero temperature.

The capability to provide a consistent theoretical description of a variety of nuclear matter properties relevant to neutron stars—including the equation of state, the Love numbers determining tidal deformation, the complex frequencies of the quasi normal oscillation modes, the transport coefficients, and the neutrino emission rates—will be needed for the interpretation of the unprecedented wealth of information expected from gravitational wave detection and multi-messenger astronomy.

Comparison to the data will be critical to assess the validity of theoretical models of nuclear dynamics, as well as to establish the limits of the paradigm underlying nuclear matter theory. Even though the observation of y-scaling in electron-nucleus scattering at high momentum transfer provides convincing evidence that the beam particles couple to nucleons carrying momenta as high as 700 MeV—corresponding to the Fermi momentum of neutron matter at density exceeding nuclear saturation density by almost a factor of ten—the description in terms of point-like nucleons is, in fact, likely to break down in the core of massive neutron stars, where the transition to new phases of matter comprising strange baryons or deconfined quarks is expected to occur.

APPENDIX A

Two- and Three-Body Cluster Contributions

In this Appendix we provide a detailed description of the two- and the three- body cluster contributions to the energy per particle of nuclear matter. The discussion will be limited to the case of static interactions, not including momentum dependent terms. In addition to being vaulable from a pedagogical point of view—most notably for the understanding of the treatment of reducible diagrams—this analysis is meant to clarify the formalism underlying the development of the CBF effective interaction, introduced in Section 5.4.

Using the notation of Section 5.3, the two-body contribution to the product $F^\dagger v_{12} F$ can be written in the form

$$F^\dagger v_{12} F \Big|_{2b} \equiv X^{(2)}(x_1, x_2) = F_{12} v_{12} F_{12} \,. \tag{A.1}$$

In what follows, we will not discuss the cluster expansion of the denominator, because its contribution is canceled exactly by the sum of the disconnected parts of the diagrams in the numerator.

The two-body contribution to the ground-state expectation value of the potential is given by

$$\begin{aligned}\langle v \rangle \Big|_{2b} &= \frac{1}{2} \sum_{\mathbf{k}_1,\mathbf{k}_2} \int dx_1 dx_2 \phi^\dagger_{\mathbf{k}_1}(x_1) \phi^\dagger_{\mathbf{k}_2}(x_2) F_{12} v_{12} F_{12} (1 - P_{12}) \phi_{\mathbf{k}1}(x_1) \phi_{\mathbf{k}_2}(x_2) \\ &= \frac{\varrho^2}{2} \int d\mathbf{r}_1 d\mathbf{r}_2 \; \mathrm{CTr}_{12} \left[F_{12} v_{12} F_{12} (1 - P_{12} \ell^2_{12}) \right] .\end{aligned} \tag{A.2}$$

where we have introduced the notation $\mathrm{CTr}_{1\ldots n} = \mathrm{CTr}_1 \ldots \mathrm{CTr}_n$ Owing to translation invariance, the integration over the centre-of-mass coordinate, $\mathbf{R}_{12} = (\mathbf{r}_1 + \mathbf{r}_2)/2$, can be trivially carried out. The resulting expression of the potential energy per nucleon is

$$\frac{\langle v \rangle}{A} \Big|_{2b} = \frac{\varrho}{2} \int d\mathbf{r}_{12} \, \mathrm{CTr}_{12} [F_{12} v_{12} F_{12} (1 - P_{12} \ell^2_{12})] \,. \tag{A.3}$$

The two-body term arising from the cluster expansion of the kinetic energy operator, $T = -\boldsymbol{\nabla}^2/2m$, reads

$$-\frac{1}{2m} \boldsymbol{\nabla}^2_1 F^\dagger F \Big|_{2b} \equiv \sum_{1<i} X^2(x_1; x_i) = \sum_{1<i}^{A} \Big(-\frac{1}{2m} F_{1i} [\boldsymbol{\nabla}^2_1, F_{1i}] \Big) \,, \tag{A.4}$$

where the commutator removes the Fermi gas energy, which is a one-body contribution. Exploiting the symmetry of the wave function we easily obtain

$$\langle T\rangle\Big|_{2b} = -\frac{1}{2m}\sum_{\mathbf{k}_1,\mathbf{k}_2}\int dx_1 dx_2 \phi^\dagger_{\mathbf{k}_1}(x_1)\phi^\dagger_{\mathbf{k}_2}(x_2)F_{12}[\nabla_1^2, F_{12}](1-P_{12})\phi_{\mathbf{k}_1}(x_1)\phi_{\mathbf{k}_2}(x_2)\,. \quad (A.5)$$

In order to eliminate the term involving gradients acting on both the correlation function, F_{12}, and the plane wave, it is convenient to perform an integration by parts, as discussed in Chapter 4, with the result

$$\langle T\rangle\Big|_{2b} = \frac{1}{2m}\sum_{\mathbf{k}_1,\mathbf{k}_2}\int dx_1 dx_2 \phi^\dagger_{\mathbf{k}_1}(x_1)\phi^\dagger_{\mathbf{k}_2}(x_2)(\nabla_1 F_{12})(\nabla_1 F_{12})(1-P_{12})\phi_{\mathbf{k}_1}(x_1)\phi_{\mathbf{k}_2}(x_2)\,. \quad (A.6)$$

Finally, because $\nabla_1 F_{12} = \nabla_{12} F_{12}$, we can perform the integration over the center-of-mass coordinate, to obtain the following expression of the kinetic energy per particle

$$\frac{\langle T\rangle}{A}\Big|_{2b} = \frac{1}{2m}\varrho\int d\mathbf{r}_{12}\,\mathrm{CTr}_{12}[(\nabla_{12}F_{12})(\nabla_{12}F_{12})(1-P_{12}\ell^2_{12})]\,. \quad (A.7)$$

The calculation of $(\nabla_1 F_{12})(\nabla_1 F_{12})$ is performed taking into account the angular dependence of the tensor operator. From the relation

$$\nabla_1 f(r_{12})\nabla_1 S_{12} = 0\,, \quad (A.8)$$

it follows that

$$\begin{aligned}(\nabla_1 F_{12})(\nabla_1 F_{12}) &= \sum_{p,q}[(\nabla_1 f^p_{12})O^p_{12} + f^p_{12}(\nabla_1 O^p_{12})][(\nabla_1 f^q_{12})O^q_{12} + f^q_{12}(\nabla_1 O^q_{12})]\\ &= \sum_{p,q}[(\nabla_1 f^p_{12})O^p_{12}(\nabla_1 f^q_{12})O^q_{12} + f^p_{12}(\nabla_1 O^p_{12})f^q_{12}(\nabla_1 O^q_{12})]\,. \quad (A.9)\end{aligned}$$

The first term in the right-hand side of Eq. (A.9) can be conveniently rewritten in terms of derivatives with respect to the magnitude of $\mathbf{r}_{12}$, with the result

$$\sum_{p,q}(\nabla_1 f^p_{12})O^p_{12}(\nabla_1 f^q_{12})O^q_{12} = \sum_{p,q} f^{p\,\prime}_{12}\,O^p_{12}\,f^{q\,\prime}_{12}\,O^q_{12}\,. \quad (A.10)$$

Moreover, using the identity

$$(\nabla_1 S_{12})(\nabla_1 S_{12}) = \frac{6}{r^2_{12}}(6 + 2\sigma_{12} + S_{12})\,, \quad (A.11)$$

with $\sigma_{12} = (\boldsymbol{\sigma}_1\cdot\boldsymbol{\sigma}_2)$, in the second term we obtain

$$\sum_{p,q}[f^p_{12}(\nabla_1 O^p_{12})f^q_{12}(\nabla_1 O^q_{12})] = \frac{6}{r^2_{12}}(f^t_{12} + f^{t\tau}_{12}\tau_{12})^2(6 + 2\sigma_{12} + S_{12}) \quad (A.12)$$

The three-body cluster contribution appearing in the expansion of $F^\dagger v_{12}F$ is given by

$$\begin{aligned}F^\dagger v_{12}F\Big|_{3b} &= \sum_{i>2} X^{(3)}(x_1, x_2; x_i)\\ &= \sum_{i>2}\left[\left(\mathcal{S}F_{12}F_{1i}F_{2i}\right)v_{12}\left(\mathcal{S}F_{12}F_{1i}F_{2i}\right) - F_{12}v_{12}F_{12}\right]. \quad (A.13)\end{aligned}$$

Within the diagrammatic scheme developed by S. Fantoni and S. Rosati [73], the three-body cluster contribution to $\langle v_{12} \rangle$ does not coincide with the expectation value of the above quantity, because the reducible diagrams arising from four-body clusters contributing to $F^\dagger v_{12} F$ must be also taken into account.

The ground-state expectation value of the direct three-body cluster contribution, given by [176]

$$\langle v_{12} \rangle \Big|_{3b}^{\text{dir}} = \frac{\varrho^2}{2} \int d\mathbf{r}_{12} d\mathbf{r}_{13} \, \mathrm{CTr}_{123} \Big[(\mathcal{S}F_{12}F_{13}F_{23}) v_{12} (\mathcal{S}F_{12}F_{13}F_{23}) - F_{12} v_{12} F_{12} (F_{13}^2 + F_{23}^2 - 1) \Big] , \tag{A.14}$$

includes both a term

$$(\mathcal{S}F_{12}F_{13}F_{23}) v_{12} (\mathcal{S}F_{12}F_{13}F_{23}) - F_{12} v_{12} F_{12} \tag{A.15}$$

belonging to the three-body cluster contribution of $F^\dagger v_{12} F$, and a term arising from the reducible four-body contribution

$$-F_{12} v_{12} F_{12} (F_{13}^2 + F_{23}^2 - 2) , \tag{A.16}$$

as illustrated in Fig.A.1.

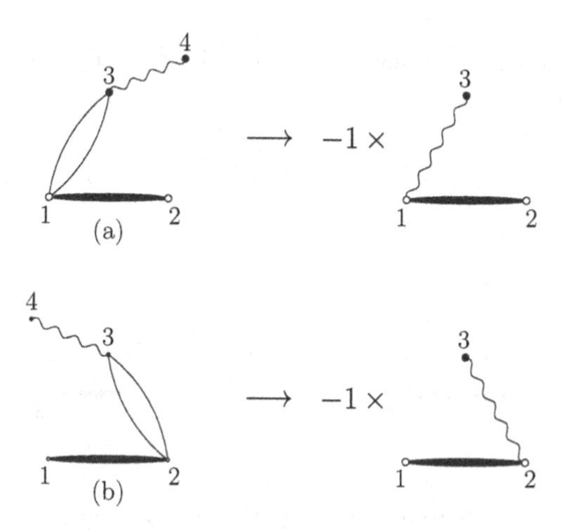

Figure A.1 Three-body reduction of four-body reducible diagrams.

It is worth pointing out that in the case of Jastrow correlations, in which $F_{ij} = f^c(r_{ij})$, the three- and four-body reducible diagrams cancel completely, as discussed in Section 5.2.1, and we are left with the irreducible contribution

$$\langle v_{12} \rangle \Big|_{3b}^{\text{dir-Jastrow}} = \frac{\varrho^2}{2} \int d\mathbf{r}_{12} d\mathbf{r}_{13} f^c(r_{12})^2 v(r_{12}) (f^c(r_{13})^2 - 1)(f^c(r_{23})^2 - 1) . \tag{A.17}$$

The sum of diagrams involving the exchange of particles 1 and 2 turns out to be

$$\langle v_{12} \rangle \Big|_{3b}^{P_{12}} = -\frac{\varrho^2}{2} \int d\mathbf{r}_{12} d\mathbf{r}_{13} \ell^2(r_{12}) \, \mathrm{CTr}_{123} \Big[(\mathcal{S}F_{12}F_{13}F_{23}) v_{12} (\mathcal{S}F_{12}F_{13}F_{23}) P_{12} - F_{12} v_{12} F_{12} (F_{13}^2 + F_{23}^2 - 1) P_{12} \Big] . \tag{A.18}$$

The four-body diagram leading to the appearance of the term $\frac{1}{2}F_{12}v_{12}F_{12}(F_{13}^2+F_{23}^2-2)P_{12}$ is shown in Fig. A.2

The diagrams in which particles 1 and 3 are exchanged give rise to the contribution

$$\langle v_{12}\rangle\Big|_{3b}^{P_{13}} = -\frac{\varrho^2}{2}\int d\mathbf{r}_{12}d\mathbf{r}_{13}\ell^2(r_{13})\ \mathrm{CTr}_{123}\Big[(\mathcal{S}F_{12}F_{13}F_{23})v_{12}(\mathcal{S}F_{12}F_{13}F_{23})P_{13} - F_{12}v_{12}F_{12}F_{13}^2P_{13}\Big]\ , \tag{A.19}$$

involving the term $\frac{1}{2}F_{12}v_{12}F_{12}(F_{13}^2-1)P_{13}$ originating from the four-body reducible diagram of Fig. A.3.

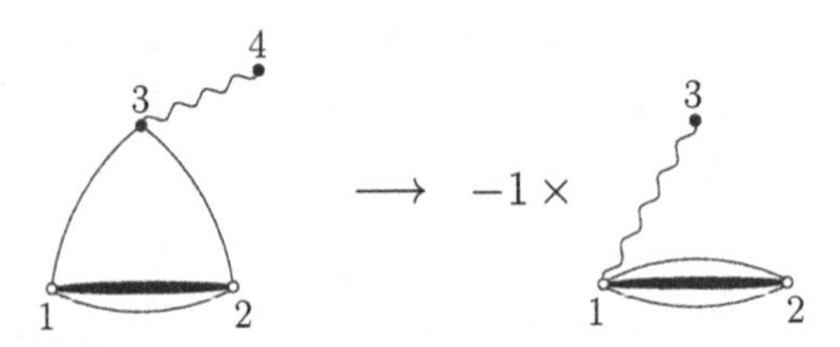

Figure A.2 Reduction of a four-body diagram, giving rise to a three-body diagram involving the exchange of particles 1 and 2.

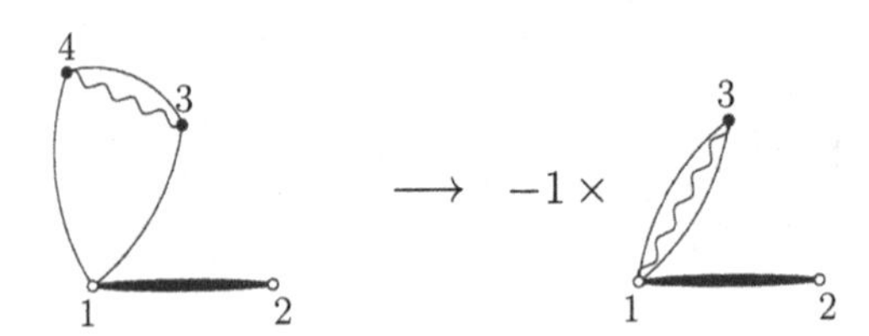

Figure A.3 Reduction of a four-body diagram, giving rise to a three-body diagram involving the exchange of particles 1 and 3.

Since the potential is invariant under the exchange $x_1 \leftrightharpoons x_2$, the contribution of diagrams involving an exchange between particles 2 and 3 is also given by Eq. (A.19). The associated four-body reducible diagram is very similar to the one of Fig. A.3 but with the loop attached to particle 2 instead of particle 1.

Finally, let us consider the diagrams with a circular exchange involving particles 1, 2 and 3. In this case there are no reducible four-body diagrams that partly cancel the reducible part of the three body diagram. In fact, there are no three-body reducible diagrams involving circular exchange at all. However, the four-body diagram of Fig. A.4, having no correlation lines connecting particles 1 and 2 to the other particles, can be reduced to a three-body diagram. The resulting three-body term, involving a circular exchange, reads

$$\langle v_{12}\rangle\Big|_{3b}^{\rm cir} = \varrho^2 \int d\mathbf{r}_{12}d\mathbf{r}_{13}\ell(r_{12})\ell(r_{13})\ell(r_{23})\ {\rm CTr}_{123}\Big[(\mathcal{S}F_{12}F_{13}F_{23})v_{12}(\mathcal{S}F_{12}F_{13}F_{23})P_{12}P_{13} - F_{12}v_{12}F_{12}F_{13}^2P_{13}P_{12}\Big]\ . \tag{A.20}$$

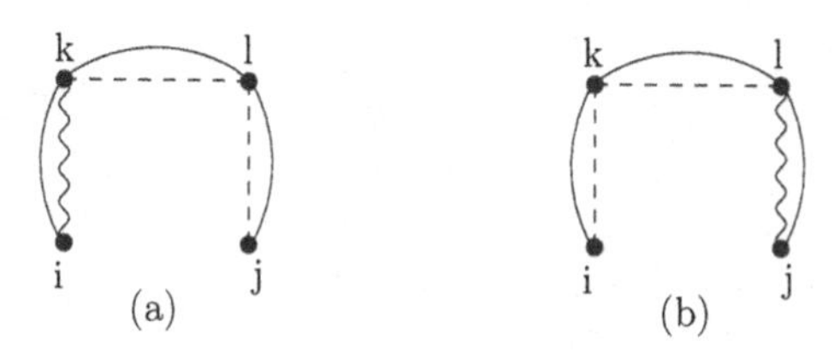

Figure A.4 Four-body diagrams contributing to the three-body diagrams with a circular exchange involving particles 1, 2 and 3.

The three-body cluster contribution to the PB form of the kinetic energy, discussed in Chapter 4, comprises terms involving $\nabla_1^2(\mathcal{S}F_{12}F_{13}F_{23})$. Their explicit expressions can be obtained from the corresponding equations for the two-body potential, by substituting

$$v_{12}(\mathcal{S}F_{12}F_{13}F_{23}) \to -2[\mathcal{S}(\boldsymbol{\nabla}_1^2F_{12})F_{13}F_{23}] - 2[\mathcal{S}(\boldsymbol{\nabla}_1F_{12})\cdot(\boldsymbol{\nabla}_1F_{13})F_{23}]\ , \tag{A.21}$$

and

$$v_{12}F_{12} \to -2(\boldsymbol{\nabla}_1^2F_{12})\ . \tag{A.22}$$

In the literature, the contributions of terms involving the operators $(\boldsymbol{\nabla}_1^2F_{12})$ and $(\boldsymbol{\nabla}_1F_{12})\cdot(\boldsymbol{\nabla}_1F_{13})$ are often referred to as $W^{\rm kin}$ and U_F, respectively, while W_F corresponds to the diagrams in which particles 1 and 2 are exchanged or belong to a circular fermion loop. Following this notation, we can write

$$\langle T\rangle\Big|_{3b\,W_F}^{P_{12}} = \frac{1}{m}\varrho^2 \int d\mathbf{r}_{12}d\mathbf{r}_{13}\ell(r_{12})\ell'(r_{12})\hat{r}_{12}\cdot{\rm CTr}_{123}\Big\{(\mathcal{S}F_{12}F_{13}F_{23}) \times[\mathcal{S}(\boldsymbol{\nabla}_1F_{12})F_{13}F_{23})P_{12} - F_{12}(\boldsymbol{\nabla}_1F_{12})(F_{13}^2+F_{23}^2-1)P_{12}]\Big\}\ , \tag{A.23}$$

where $\hat{r}_{12} = \mathbf{r}_{12}/|\mathbf{r}_{12}|$, and

$$\langle \hat{T}\rangle\Big|_{3b\,W_F}^{\rm cir} = -\frac{1}{m}\varrho^2 \int d\mathbf{r}_{12}d\mathbf{r}_{13}\ell(r_{13})\ell(r_{23})\ell'(r_{12})\hat{r}_{12}\cdot{\rm CTr}_{123}\Big\{(\mathcal{S}F_{12}F_{13}F_{23}) \times[\mathcal{S}(\boldsymbol{\nabla}_1F_{12})F_{13}F_{23}]P_{12}P_{13} - F_{12}(\boldsymbol{\nabla}_1F_{12})F_{13}^2P_{13}P_{12}]\Big\}\ . \tag{A.24}$$

The contributions to U_F correspond to diagrams involving the exchange operator P_{13}

$$\langle \hat{T}\rangle\Big|_{3b\,U_F}^{P_{13}} = \frac{1}{m}\varrho^2 \int d\mathbf{r}_{12}d\mathbf{r}_{13}\ell'(r_{13})\hat{r}_{13}\cdot{\rm CTr}_{123}\Big[(\mathcal{S}F_{12}F_{13}F_{23}) \times\mathcal{S}(\boldsymbol{\nabla}_1F_{12})F_{13}F_{23})P_{13}\Big]\ , \tag{A.25}$$

or a circular exchange

$$\langle T\rangle\Big|_{3b\,U_F}^{\mathrm{cir}} = -\frac{\hbar^2}{m}\varrho^2 \int d\mathbf{r}_{12}d\mathbf{r}_{13}\ell(r_{12})\ell(r_{13})\ell'(r_{13})\hat{r}_{23}\cdot \mathrm{CTr}_{123}\Big[(\mathcal{S}F_{12}F_{13}F_{23}) \times [\mathcal{S}(\boldsymbol{\nabla}_1 F_{12})F_{13}F_{23}]P_{12}P_{13}\Big]\,. \qquad \text{(A.26)}$$

Note that in this case there are no subtraction terms arising from reducible diagrams.

Bibliography

[1] G. Gamow, "Mass defect curve and nuclear constitution," *Proc. R. Soc. Lond. A*, vol. 126, p. 632, 1930.

[2] C. v. Weizsäcker, "Zur theorie der kernmassen," *Zeitschrift für Physik A*, vol. 96, p. 431, 1935.

[3] L.D. Landau, "On the theory of stars," *Phys. Z. Sowjetunion*, vol. 1, p. 285, 1932.

[4] W. Baade and F. Zwicky, "On super-novae," *Proc Natl. Acad. Sci. USA*, vol. 20, p. 254, 1934.

[5] A. Hewish, S.J. Bell, J.D.H. Pilkington, P.F. Scott, and R.A. Collins, "Observation of a rapidly pulsating radio source," *Nature*, vol. 217, p. 709, 1968.

[6] J.R. Oppenheimer and G.M. Volkoff, "On massive neutron cores," *Phys. Rev.*, vol. 55, p. 374, 1939.

[7] I. Sick, "Proton charge radius from electron scattering," *Atoms*, vol. 6, p. 2, 2017.

[8] R. Bradford, A. Bodek, H. Budd, and J. Arrington, "A new parameterization of the nucleon elastic form factors," *Nucl. Phys. B Proc. Suppl.*, vol. 159, p. 127, 2006.

[9] J. Arrington *et al.*, "Inclusive electron-nucleus scattering at large momentum transfer," *Phys. Rev. Lett.*, vol. 82, p. 2056, 1999.

[10] H. Yukawa, "On the interaction of elementary particles," *Proc. Phys. Math. Soc. Jpn.*, vol. 17, p. 48, 1935.

[11] N. Ishii, S. Aoki, and T. Hatsuda, "Nuclear force from lattice QCD," *Phys. Rev. Lett.*, vol. 99, p. 022001, 2007.

[12] J. Hu, H. Toki, and H. Shen, "The properties of nuclear matter with lattice NN potential in relativistic Brueckner-Hartree-Fock theory," *Scientific Reports*, vol. 6, p. 35590, 2016.

[13] R. B. Wiringa, V. G. J. Stoks, and R. Schiavilla, "Accurate nucleon-nucleon potential with charge-independence breaking," *Phys. Rev. C*, vol. 51, p. 38, 1995.

[14] R. B. Wiringa and S. C. Pieper, "Evolution of nuclear spectra with nuclear forces," *Phys. Rev. Lett.*, vol. 89, p. 182501, 2002.

[15] V.G.J. Stoks, R. A. M. Klomp, M.C.M. Rentmeester, and J.J. de Swart, "Partial-wave analysis of all nucleon-nucleon scattering data below 350 MeV," *Phys. Rev. C*, vol. 48, p. 792, 1993.

[16] J.R. Bergervoet *et al.*, "Phase shift analysis of all proton-proton scattering data below T_{lab}=350 MeV," *Phys. Rev. C*, vol. 41, p. 1435, 1990.

[17] R.L. Workman, W.J. Briscoe, and I.I. Strakovsky, "Partial-wave analysis of nucleon-nucleon elastic scattering data," *Phys. Rev. C*, vol. 94, p. 065203, 2016.

[18] J. L. Friar, "Trinucleon bound states," in *New Vistas in Electro-Nuclear Physics* (E. L. Tomusiak, H. S. Caplan, and E. T. Dressler, eds.), p. 213, Plenum Press, New York, 1986.

[19] J. Fujita and H. Miyazawa, "Pion theory of three-body forces I," *Prog. Theor. Phys.*, vol. 17, p. 360, 1957.

[20] B. S. Pudliner, V. R. Pandharipande, J. Carlson, and R. B. Wiringa, "Quantum Monte Carlo calculations of $A \leq 6$ nuclei," *Phys. Rev. Lett.*, vol. 74, p. 4396, 1995.

[21] S. C. Pieper, "The Illinois extension to the Fujita-Miyazawa three-nucleon force," *AIP Conf. Proc.*, vol. 1011, p. 143, 2008.

[22] J. Carlson, S. Gandolfi, F. Pederiva, S. C. Pieper, R. Schiavilla, K. E. Schmidt, and R. B. Wiringa, "Quantum Monte Carlo methods for nuclear physics," *Rev. Mod. Phys.*, vol. 87, p. 1067, 2015.

[23] S. Bacca and S. Pastore, "Electromagnetic reactions on light nuclei," *J. Phys. G: Nucl. Part. Phys.*, vol. 41, p. 123002, 2014.

[24] S. A. Coon and H. K. Ha, "Reworking the Tucson-Melbourne three-nucleon potential," *Few Body Syst.*, vol. 30, p. 101, 2001.

[25] S. C. Pieper and R. B. Wiringa, "Quantum Monte Carlo calculation of light nuclei," *Annu. Rev. Nucl. Part. Sci.*, vol. 51, p. 53, 2001.

[26] R. Machleidt, K. Holinde and Ch. Elster, "The Bonn meson-exchange model for the nucleon-nucleon interaction," *Phys. Rep.*, vol. 149, p. 1, 1987.

[27] R. Machleidt, "High-precision, charge-dependent Bonn nucleon-nucleon potential," *Phys. Rev. C*, vol. 63, p. 024001, 2001.

[28] S. Weinberg, "Nuclear forces from chiral lagrangians," *Phys. Lett. B*, vol. 251, p. 288, 1990.

[29] E. Epelbaum, H.-W. Hammer, and U.-G. Meißner, "Modern theory of nuclear forces," *Rev. Mod. Phys.*, vol. 81, p. 1773, 2009.

[30] R. Machleidt and D. Entem, "Chiral effective field theory and nuclear forces," *Phys. Rep.*, vol. 503, p. 1, 2011.

[31] A. Gezerlis, I. Tews, E. Epelbaum, S. Gandolfi, K. Hebeler, A. Nogga, and A. Schwenk, "Quantum Monte Carlo calculations with chiral effective field theory interactions," *Phys. Rev. Lett.*, vol. 111, p. 032501, 2013.

[32] A. Gezerlis, I. Tews, E. Epelbaum, M. Freunek, S. Gandolfi, K. Hebeler, A. Nogga, and A. Schwenk, "Local chiral effective field theory interactions and quantum Monte Carlo applications," *Phys. Rev. C*, vol. 90, p. 054323, 2014.

[33] D. Lonardoni, J. Carlson, S. Gandolfi, J. E. Lynn, K. E. Schmidt, A. Schwenk, and X. B. Wang, "Properties of nuclei up to $A = 16$ using local chiral interactions," *Phys. Rev. Lett.*, vol. 120, p. 122502, 2018.

[34] M. Piarulli, A. Baroni, L. Girlanda, A. Kievsky, A. Lovato, E. Lusk, L. E. Marcucci, S. C. Pieper, R. Schiavilla, M. Viviani, and R. B. Wiringa, "Light-nuclei spectra from chiral dynamics," *Phys. Rev. Lett.*, vol. 120, p. 052503, 2018.

[35] D. Logoteta, I. Bombaci, and A. Kievsky, "Nuclear matter properties from local chiral interactions with Δ isobar intermediate states," *Phys. Rev. C*, vol. 94, p. 064001, 2016.

[36] J.E. Lynn, I. Tews, J. Carlson,S. Gandolfi, A. Gezerlis, K.E. Schmidt, and A. Schwenk, "Chiral three-nucleon interactions in light nuclei, neutron-α scattering, and neutron matter," *Phys. Rev. Lett.*, vol. 116, p. 062501, 2016.

[37] O. Benhar, "Scale Dependence of Nucleon-Nucleon Potentials." arXiv:1903.11353 [nucl-th], 2019.

[38] M. Piarulli, L. Girlanda, R. Schiavilla, R. N. Pérez, J. E. Amaro, and E. R. Arriola, "Minimally nonlocal nucleon-nucleon potentials with chiral two-pion exchange including Δ resonances," *Phys. Rev. C*, vol. 91, p. 024003, 2015.

[39] K. Huang, *Statistical sechanics.* John Wiley & Sons, New York, New York, 1963.

[40] L.D. Landau and E. M. Lifshitz, *Statistical Physics, Part 1.* Pergamon Press, Oxford, 1980.

[41] S. Shlomo, V. M. Kolomietz, and G. Colò, "Deducing the nuclear-matter incompressibility coefficient from data on isoscalar compression modes," *The European Physical Journal A - Hadrons and Nuclei*, vol. 30, no. 1, p. 23, 2006.

[42] G. Colò, "The compression modes in atomic nuclei and their relevance for the nuclear equation of state," *Physics of Particles and Nuclei*, vol. 39, no. 2, p. 286, 2008.

[43] Ya. B. Zel'dovich, "The equation of state at ultrahigh densities and its relativistic limitations," *Sov. Phys. JETP*, vol. 14, p. 1143, 1962.

[44] B.-A. Li and X. Han, "Constraining the neutron-proton effective mass splitting using empirical constraints on the density dependence of nuclear symmetry energy around normal density," *Phys. Lett. B*, vol. 727, p. 276, 2013.

[45] P. Danielewicz, R. Lacey, and W. G. Lynch, "Determination of the equation of state of dense matter," *Science*, vol. 298, p. 1592, 2002.

[46] G. Baym and C. J. Pethick, *Landau Fermi-Liquid Theory.* John Wiley & Sons, New York, 1991.

[47] N. M. Hugenholtz and L. Van Hove, "A theorem on the single particle energy in a Fermi gas with interaction," *Physica*, vol. 24, p. 363, 1958.

[48] V.M. Galitskii and A.B. Migdal, "Application of quantum field theory methods to the many-body problem," *Sov. Phys. JETP.*, vol. 34, p. 96, 1958.

[49] H. A. Bethe, "Theory of nuclear matter," *Ann. Rev. Nucl. Sci.*, vol. 21, p. 93, 1971.

[50] O. Benhar, "Exploring nuclear dynamics with $(e, e'p)$ reactions: from LNF to JLab," *Nucl. Phys. News*, vol. 26, p. 15, 2016.

[51] J. M. Blatt and V. F. Weisskopf, *Theoretical Nuclear Physics.* Dover Publications, New York, 1991.

[52] E. Feenberg, *Theory of Quantum Fluids.* Academic Press, New York, 1969.

[53] O. Benhar, A. Fabrocini, and S. Fantoni, "Occupation probabilities and hole-state strengths in nuclear matter," *Phys. Rev. C*, vol. 41, p. R24, 1990.

[54] B.D. Day, "Elements of the Brueckner-Goldstone theory of nuclear matter," *Rev. Mod. Phys.*, vol. 39, p. 719, 1967.

[55] J.W. Clark, "Variational theory of nuclear matter," *Prog. Part. Nucl. Phys.*, vol. 2, p. 89, 1979.

[56] V.R. Pandharipande and R.B. Wiringa, "Variations on a theme of nuclear matter," *Rev. Mod. Phys.*, vol. 51, p. 821, 1979.

[57] K. Hebeler and A. Schwenk, "Chiral three-nucleon forces and neutron matter," *Phys. Rev. C*, vol. 82, p. 014314, 2010.

[58] K. Hebeler and R. J. Furnstahl, "Neutron matter based on consistently evolved chiral three-nucleon interactions," *Phys. Rev. C*, vol. 87, p. 031302, 2013.

[59] C. Drischler, K. Hebeler, and A. Schwenk, "Asymmetric nuclear matter based on chiral two- and three-nucleon interactions," *Phys. Rev. C*, vol. 93, p. 054314, 2016.

[60] O. Benhar and A. Lovato, "Perturbation theory of nuclear matter with a microscopic effective interaction," *Phys. Rev. C*, vol. 96, p. 054301, 2017.

[61] J.P. Jeukenne, A. Lejeunne, and C. Mahaux, "Many-body theory of nuclear matter," *Phys. Rep.*, vol. 25, p. 83, 1976.

[62] J. Goldstone, "Derivation of the Brueckner many-body theory," *Proc. Roy. Soc.*, vol. 239, p. 267, 1957.

[63] B.H. Brandow, "Compact-cluster expansion for the nuclear many-body problem," *Phys. Rev.*, vol. 152, p. 863, 1966.

[64] P.J. Siemens, "Nuclear-matter reaction matrix," *Nucl. Phys. A*, vol. 141, p. 225, 1970.

[65] B.D. Day, "Current state of nuclear matter calculations," *Rev. Mod. Phys.*, vol. 50, p. 495, 1978.

[66] M. Baldo, A. Fiasconaro, H.Q. Song, G. Giansiracusa, and U. Lombardo, "High density symmetric nuclear matter in the Bethe-Brueckner-Goldstone approach," *Phys. Rev. C*, vol. 65, p. 017303, 2001.

[67] R. Rajaraman and H.A. Bethe, "Three-body problem in nuclear matter," *Rev. Mod. Phys.*, vol. 39, p. 745, 1967.

[68] F. Coester, S. Cohen, B.D. Day, and C.M. Vincent, "Variation in nuclear-matter binding energies with phase-shift-equivalent two-body potentials," *Phys. Rev. C*, vol. 1, p. 769, 1970.

[69] X.R. Zhou, G.F. Burgio, U. Lombardo, H.-J. Schulze, and W. Zuo, "Three-body forces and neutron star structure," *Phys. Rev. C*, vol. 69, p. 018801, 2004.

[70] P. Grangé, A. Lejeune, M. Martzolff, and J.-F. Mathiot, "Consistent three-nucleon forces in the nuclear many-body problem," *Phys. Rev. C*, vol. 40, p. 1040, 1989.

[71] R. Jastrow, "Many-body problem with strong forces," *Phys. Rev.*, vol. 98, p. 1479, 1955.

[72] S.A. Rice and P. Gray, *Statistical Mechanics of Simple Liquids.* Interscience, New York, 1966.

[73] S. Fantoni and S. Rosati, "Jastrow correlations and an irreducible cluster expansion for infinite boson or fermion systems," *Nuovo Cim. A*, vol. 20, p. 179, 1974.

[74] S. Fantoni and S. Rosati, "Calculation of the two-body correlation function for fermion systems," *Lett. Nuovo Cim.*, vol. 10, p. 545, 1974.

[75] M. Gaudin, J. Gillespie, and G. Ripka, "Jastrow correlations," *Nucl. Phys. A*, vol. 176, p. 179, 1971.

[76] V.R. Pandharipande and H.A. Bethe, "Variational method for dense systems," *Phys. Rev. C*, vol. 7, p. 1312, 1973.

[77] J.W. Clark and P. Westhaus, "Method of correlated basis functions," *Phys. Rev.*, vol. 141, p. 833, 1966.

[78] S. Fantoni and S. Rosati, "The Jackson-Fennberg form of the kinetic energy in the chain and FHNC approximations," *Phys. Lett. B*, vol. 84, p. 23, 1979.

[79] O. Benhar, C. Ciofi degli Atti, S. Fantoni, S. Rosati, A. Kallio, L. Lantto, and P. Toropainen, "Lowest-order and hyper-netted-chain calculations of nuclear matter," *Phys. Lett. B*, vol. 64, p. 395, 1976.

[80] V.R. Pandharipance, R.B. Wiringa, and B.D. Day, "Do lowest-order approximations adequately describe nuclear matter?," *Phys. Lett. B*, vol. 57, p. 205, 1975.

[81] F. Coester and H. Kümmel, "Short-range correlations in nuclear wave functions," *Nucl. Phys.*, vol. 17, p. 477, 1960.

[82] J. Lindgren abd J. Morrison, *Atomic Many-Body Theory.* Springer, Berlin, 1985.

[83] B.D. Day and J.G. Zabolitzky, "Coupled-cluster calculation for nuclear matter and comparison with the hole-line expansion," *Nucl. Phys. A*, vol. 366, p. 221, 1981.

[84] D.R. Entem abd R. Machleidt, "Accurate charge-dependent nucleon-nucleon potential at fourth order of chiral perturbation theory," *Phys. Rev. C*, vol. 68, p. 041001(R), 2003.

[85] G. Baardsen, A. Ekström, G. Hagen, and M. Hjorth-Jensen, "Coupled-cluster studies of infinite nuclear matter," *Phys. Rev. C*, vol. 88, p. 054312, 2013.

[86] G. Hagen, T. Papenbrock, A. Ekström, K. A. Wendt, G. Baardsen, S. Gandolfi, M. Hjorth-Jensen, and C. J. Horowitz, "Coupled-cluster calculations of nucleonic matter," *Phys. Rev. C*, vol. 89, p. 014319, 2014.

[87] G. Hagen, T. Papenbrock, M. Hjorth-Jensen, D. J. Dean, "Coupled-cluster computations of atomic nuclei," *Rep. Prog. Phys.*, vol. 096302, p. 1, 2014.

[88] A. Ramos, A. Polls, and W.H. Dickhoff, "Single-particle properties and short-range correlations in nuclear matter," *Nucl. Phys. A*, vol. 503, p. 1, 1989.

[89] C. Barbieri and A. Carbone, "Self-consistent Green's function approaches," in *An Advanced Course in Computational Nuclear Physics: Bridging the Scales from Quarks to Neutron Stars* (M. Hjorth-Jensen, M. P. Lombardo, and U. van Kolck, ed.), p. 571, Springer, Berlin, 2017.

[90] A. Carbone, A. Rios, and A. Polls, "Symmetric nuclear matter with chiral three-nucleon forces in the self-consistent Green's functions approach," *Phys. Rev. C*, vol. 88, p. 044301, 2013.

[91] M.H. Kalos, *Monte Carlo Methods in Quantum Problems.* NATO ASI Series: Mathematical and Physical Sciences, D. Reidel Publishing Company, Dordrecht, 1984.

[92] M.H. Kalos and P.A. Whitlock, *Monte Carlo Methods.* Wiley-VCH Verlag GmbH, Weinheim, 2008.

[93] J. Carlson, J. Morales, Jr., V.R. Pandharipande, and D.G. Ravenhall, "Quantum Monte Carlo calculations of neutron matter," *Phys. Rev. C*, vol. 68, p. 025802, 2003.

[94] M.H. Kalos, "Monte Carlo calculations of the ground state of three- and four-body nuclei," *Phys. Rev.*, vol. 128, p. 1791, 1962.

[95] S. Fantoni and K.E. Schmidt, "A quantum Monte Carlo method for nucleon systems," *Phys. Lett. B*, vol. 446, pp. 99 – 103, 1999.

[96] S. Gandolfi, F. Pederiva, S. Fantoni, and K.E. Schmidt, "Quantum Monte Carlo calculations of symmetric nuclear matter," *Phys. Rev. Lett.*, vol. 98, p. 102503, 2007.

[97] S. Gandolfi, A. Lovato, J. Carlson,and K.E. Schmidt, "From the lightest nuclei to the equation of state of asymmetric nuclear matter with realistic nuclear interactions," *Phys. Rev. C*, vol. 90, p. 061306(R), 2014.

[98] S. Gandolfi, J. Carlson, S. Reddy, A.W. Steiner, and R. B. Wiringa, "The equation of state of neutron matter, symmetry energy and neutron star structure," *Eur. Phys. J. A*, vol. 50, p. 10, 2014.

[99] I. Tews, J. Carlson, S. Gandolfi, and S. Reddy, "Constraining the speed of sound inside neutron stars with chiral effective field theory interactions and observations," *ApJ*, vol. 860, p. 149, 2018.

[100] R.A. Krajcik and L.L. Foldy, "Relativistic center-of-mass variables for composite systems with arbitrary internal interactions," *Phys. Rev. D*, vol. 10, p. 1777, 1974.

[101] J.I. Forest, V.R. Pandharipande, and J.L. Friar, "Relativistic nuclear hamiltonians," *Phys. Rev. C*, vol. 52, p. 568, 1995.

[102] J. Carlson, V.R. Pandharipande, and R. Schiavilla, "Variational Monte Carlo calculations of ^{3}H and ^{4}He with a relativistic hamiltonian," *Phys. Rev. C*, vol. 47, p. 484, 1993.

[103] L. G. Arnold, B. C. Clark, and R. L. Mercer, "Relativistic optical model analysis of medium energy $p-^4He$ elastic scattering experiments," *Phys. Rev. C*, vol. 19, p. 917, 1979.

[104] M.R. Anastasio, L.S. Celenza, W.S. Pong, and C.M. Shakin, "Relativistic nuclear structure physics," *Phys. Rep.*, vol. 100, p. 327, 1983.

[105] R. Brockmann and R. Machleidt, "The Dirac-Brueckner Approach," in *Nuclear Methods and the Nuclear Equation of State* (M. Baldo, ed.), p. 121, World Scientific, Singapore, 1999.

[106] R. Brockmann and R. Machleidt, "Relativistic nuclear structure I. nuclear matter," *Phys. Rev. C*, vol. 42, p. 1965, 1990.

[107] F. Sammarruca, B. Chen, L. Coraggio, N. Itaco, and R. Machleidt, "Dirac-Brueckner-Hartree-Fock versus chiral effective field theory," *Phys. Rev. C*, vol. 86, p. 054317, 2012.

[108] J.D. Walecka, "A theory of highly condensed matter," *Ann. Phys. (N.Y.)*, vol. 83, p. 491, 1974.

[109] S. Fantoni and A. Fabrocini, "Correlated Basis Function Theory for Fermion Systems," in *Microscopic Quantum Many-Body Theories and Their Applications* (Jesùs Navarro and Artur Polls, ed.), vol. 510, p. 119, Springer, Berlin, 1998.

[110] S. Fantoni, "Linked-cluster perturbative expansion in correlated-basis theory," *Phys. Rev. B*, vol. 29, p. 2544, 1984.

[111] P-O. Löwdin, "On the non-orthogonality problem connected with the use of atomic wave functions in the theory of molecules and crystals," *The Journal of Chemical Physics*, vol. 18, p. 365, 1950.

[112] S. Fantoni and V.R. Pandharipande, "Orthogonalization of correlated states," *Phys. Rev. C*, vol. 37, p. 1697, 1988.

[113] J.M.J. van Leeuwen, J. Groeneveld, and J. de Boer, "New method for the calculation of the pair correlation function I," *Physica*, vol. 25, p. 792, 1959.

[114] J.P. Hansen and I.R. McDonald, *Theory of Simple Liquids.* Academic Press, New York, 2013.

[115] S. Fantoni and S. Rosati, "The hypernetted-chain approximation for a fermion system," *Nuovo Cim. A*, vol. 25, p. 593, 1975.

[116] E. Krotscheck and M.L. Ristig, "Hypernetted-chain approximation for dense Fermi fluids," *Phys. Lett. A*, vol. 48, p. 17, 1974.

[117] F. A. de Saavedra, C. Bisconti, G. Cò, and A. Fabrocini, "Renormalized fermi hypernetted chain approach in medium-heavy nuclei," *Phy. Rep.*, vol. 450, p. 1, 2007.

[118] R. Wiringa and V. Pandharipande, "A variational theory of nuclear matter (III)," *Nucl. Phys. A*, vol. 317, p. 1, 1979.

[119] R. B. Wiringa, V. Fiks, and A. Fabrocini, "Equation of state for dense nucleon matter," *Phys. Rev. C*, vol. 38, p. 1010, 1988.

[120] A. Akmal and V.R. Pandharipande, "Spin-isospin structure and pion condensation in nucleon matter," *Phys. Rev. C*, vol. 56, p. 2261, 1997.

[121] A. Akmal, V.R. Pandharipande, and D.G. Ravenhall, "Equation of state of nucleon matter and neutron star structure," *Phys. Rev. C*, vol. 58, p. 1804, 1998.

[122] B. Friedman and V. Pandharipande, "Hot and cold, nuclear and neutron matter," *Nucl. Phys. A*, vol. 361, p. 502, 1981.

[123] O. Benhar, A. Fabrocini, and S. Fantoni, "The nucleon spectral function in nuclear matter," *Nucl. Phys. A*, vol. 505, p. 267, 1989.

[124] O. Benhar, A. Fabrocini, and S. Fantoni, "Nuclear-matter green functions in correlated-basis theory," *Nucl. Phys. A*, vol. 550, p. 201, 1992.

[125] O. Benhar, D. Day, and I. Sick, "Inclusive quasielastic electron-nucleus scattering," *Rev. Mod. Phys.*, vol. 80, p. 189, 2008.

[126] S. T. Cowell and V. R. Pandharipande, "Neutrino mean free paths in cold symmetric nuclear matter," *Phys. Rev. C*, vol. 70, p. 035801, 2004.

[127] S. T. Cowell and V. R. Pandharipande, "Quenching of weak interactions in nucleon matter," *Phys. Rev. C*, vol. 67, p. 035504, 2003.

[128] A. Lovato, C. Losa, and O. Benhar, "Weak response of cold symmetric nuclear matter at three-body cluster level," *Nucl. Phys.*, vol. A901, p. 22, 2013.

[129] A. Lovato, O. Benhar, S. Gandolfi, and C. Losa, "Neutral current interactions of low-energy neutrinos in dense neutron matter," *Phys. Rev. C*, vol. 89, p. 025804, 2014.

[130] A. A. Woosley, A. Heger, and T. A. Weaver, "The evolution and explosion of massive stars," *Rev. Mod. Phys.*, vol. 74, p. 1015, 2002.

[131] S. Chandrasekhar, "The maximum mass of ideal white dwarfs," *ApJ*, vol. 74, p. 81, 1931.

[132] D. G. Ravenhall, C. J. Phethick, and J. R. Wilson, "Structure of matter below nuclear saturation density," *Phys. Rev. Lett.*, vol. 50, p. 2066, 1983.

[133] M. Hashimoto, H. Seki, and M. Yamada, "Shape of nuclei in the crust of neutron star," *Prog. Theor. Phys.*, vol. 71, p. 320, 1984.

[134] Y. Leung, *Physics of Dense Matter.* World Scientific, Singapore, 1984.

[135] M. Okamoto,T. Maruyama, K. Yabana,and T. Tatsumi, "Nuclear "pasta" structures in low-density nuclear matter and properties of the neutron-star crust," *Phys. Rev. C*, vol. 88, p. 025801, 2013.

[136] S. Weinberg, *Gravitation and Cosmology.* John Wiley & Sons, New York, 1972.

[137] R.C. Tolman, "Static solutions of Einstein's field equations for spheres of fluid," *Phys. Rev.*, vol. 55, p. 364, 1939.

[138] B.K. Harrison, K.S. Thorne, M. Wakano, and J.. Wheeler, *Gravitation theory and gravitational collapse.* University of Chicago Press, Chicago, 1965.

[139] S.L. Shapiro and S.A. Teukolsky, *Black Holes, White Dwarfs, and Neutron Stars: The Physics of Compact Objects.* John Wiley & Sons, New York, 1985.

[140] T. Damour and N. Deruelle, "General relativistic celestial mechanics of binary systems II. The post-Newtonian timing formula," *Ann. Inst. Henri Poincaré, Phys. Theor.*, vol. 4, p. 263, 1986.

[141] R.A. Hulse and J.H. Taylor, "Discovery of a pulsar in a binary system," *ApJ Lett.*, vol. 195, p. L51, 1975.

[142] J.H. Tayolr, L.A. Flower, and P.M. McCulloch, "Measurements of general relativistic effects in the binary pulsar PSR1913+16," *Nature*, vol. 277, p. 437, 1979.

[143] B. Kiziltan, A. Kottas, M. De Yoreo, and S.E. Thorsett, "The neutron star mass distribution," *ApJ*, vol. 778, p. 66, 2013.

[144] P. Demorest, T. Pennucci, S. Ransom, M. Roberts, and J. Hessels, "A two-solar-mass neutron star measured using Shapiro delay," *Nature*, vol. 467, p. 1081, 2010.

[145] J. Antoniadis *et al.*, "A massive pulsar in a compact relativistic binary," *Science*, vol. 340, p. 6131, 2013.

[146] J. Cottam, F. Paerels, and M. Mendez , "Gravitationally redshifted absorption lines in the X-ray burst spectra of a neutron star," *Nature*, vol. 420, p. 51, 2002.

[147] K. Hebeler, J. M. Lattimer, C. J. Pethick, and A. Schwenk, "Constraints on neutron star radii based on chiral effective field theory interactions," *Phys. Rev. Lett.*, vol. 105, p. 161102, 2018.

[148] G. Gamow and M. Shoenberg, "The possible role of neutrinos in stellar evolution," *Phys. Rev,*, vol. 58, p. 1117, 1940.

[149] G. Gamow and M. Shoenberg, "Neutrino theory of stellar collapse," *Phys. Rev,*, vol. 59, p. 539, 1941.

[150] C. J. Pethick, "Cooling of neutron stars," *Rev. Mod. Phys.*, vol. 64, p. 1133, 1992.

[151] G. Gamov, *My World Line: An Informal Autobiography.* Viking Press, New York, 1970.

[152] D. Page, M. Prakash, J.M. Lattimer, and A.W. Steiner, "Superfluid neutrons in the core of the neutron star in Cassiopeia A," *Proceedings of Science*, vol. XXXIVBWNP, p. 005, 2011.

[153] C.O. Heinke and A.C.G. Ho, "Direct observation of the cooling of the Cassiopeia A neutron star," *ApJ*, vol. 719, p. L167, 2010.

[154] P.S. Shternin, D.G. Yakovlev, C.O. Heinke, A.C.G. Ho, and D.J. Patnaude, "Cooling neutron star in the Cassiopeia A supernova remnant: evidence for superfluidity in the core," *MNRAS*, vol. 412, p. L108, 2011.

[155] D. Page, M. Prakash, J.M. Lattimer, and A.W. Steiner, "Rapid cooling of the neutron star in Cassiopeia A triggered by neutron superfluidity in dense matter," *Phys. Rev. Lett.*, vol. 106, p. 081101, 2011.

[156] B. P. Abbott *et al.* (LIGO Scientific Collaboration and Virgo Collaboration), "GW170817: Observation of gravitational waves from a binary neutron star inspiral," *Phys. Rev. Lett.*, vol. 119, p. 161101, 2017.

[157] B. P. Abbott *et al.* (LIGO Scientific Collaboration and Virgo Collaboration), "Multi-messenger observations of a binary neutron star merger," *The Astrophysical Journal*, vol. 848, p. L12, 2017.

[158] L. Baiotti, "Gravitational waves from neutron star mergers and their relation to the nuclear equation of state," *Prog. Part. Nucl. Phys.*, vol. 109, p. 103714, 2019.

[159] I. Tews, J. Margueron, and S. Reddy, "Critical examination of constraints on the equation of state of dense matter obtained from GW170817," *Phys. Rev. C*, vol. 98, p. 045804, 2018.

[160] P.G. Krastev and Bao-An Li, "Imprints of the nuclear symmetry energy on the tidal deformability of neutron stars," *J. Phys. G: Nucl Part. Phys.*, vol. 46, p. 074001, 2019.

[161] A. Sabatucci and O. Benhar, "Tidal deformation of neutron stars from microscopic models of nuclear dynamics." arXiv:2001.06294 [nucl-th], 2020.

[162] T. Hinderer, B. D. Lackey, R. N. Lang, and J. S. Read, "Tidal deformability of neutron stars with realistic equations of state and their gravitational wave signatures in binary inspiral," *Phys. Rev. D*, vol. 81, p. 123016, 2010.

[163] N. K. Glendenning and S. A. Moszkowski, "Reconciliation of neutron-star masses and binding of the Lambda in hypernuclei," *Phys. Rev. Lett.*, vol. 67, p. 2414, 1991.

[164] J. M. Lattimer and F. D. Swesty, "A generalized equation of state for hot, dense matter," *Nucl. Phys. A*, vol. 535, p. 331, 1991.

[165] O. Benhar, E. Berti, and V. Ferrari, "The imprint of the equation of state on the axial w-modes of oscillating neutron stars," *MNRAS*, vol. 310, p. 797, 1999.

[166] N. Andersson and K.D. Kokkotas,, "Towards gravitational wave asteroseismology," *MNRAS*, vol. 299, p. 1059, 1998.

[167] O. Benhar, V. Ferrari, and L. Gualtieri, "Gravitational wave asteroseismology reexamined," *Phys. Rev. D*, vol. 70, p. 124015, 2004.

[168] S. Chandrasekhar, "Solutions of two problems in the theory of gravitational radiation," *Phys. Rev. Lett.*, vol. 24, p. 611, 1970.

[169] S. Chandrasekhar, "The effect of gravitational radiation on the secular stability of the maclaurin spheroid," *ApJ*, vol. 161, p. 561, 1970.

[170] J.L. Friedman e B.F. Schutz, "Secular instability of rotating Newtonian stars," *ApJ*, vol. 222, p. 281, 1978.

[171] N. Andersson, "A new class of unstable modes of rotating relativistic stars," *ApJ*, vol. 502, p. 708, 1998.

[172] J.L. Friedman e S.M. Morsink, "Axial instability of rotating relativistic stars," *ApJ*, vol. 502, p. 714, 1998.

[173] K. Glampedakis and L. Gualtieri, "Gravitational waves from single neutron stars: an advanced detector era survey," in *The Physics and Astrophysics of Neutron Stars* (L. Rezzolla, P. Pizzochero, D.I. Jones, N. Rea, and I. Vidana, ed.), p. 673, Springer, Berlin, 2018.

[174] O. Benhar and M. Valli, "Shear viscosity of neutron matter from realistic nucleon-nucleon interactions," *Phys. Rev. Lett.*, vol. 99, p. 232501, 2007.

[175] O. Benhar, A. Polls, M. Valli, and I. Vidaña, "Microscopic calculations of transport properties of neutron matter," *Phys. Rev. C*, vol. 81, p. 024305, 2010.

[176] J. Morales, V. R. Pandharipande, and D. G. Ravenhall, "Improved variational calculations of nucleon matter," *Phys. Rev. C*, vol. 66, p. 054308, 2002.

Index